Flying
the
Mountains

Flying the Mountains

A Training Manual for Flying Single-Engine Aircraft

Fletcher Anderson

McGraw-Hill

New York   Chicago   San Francisco   Lisbon   London   Madrid
Mexico City   Milan   New Delhi   San Juan   Seoul
Singapore   Sydney   Toronto

The **McGraw·Hill** Companies

Cataloging-in-Publication Data is on file with the Library of Congress

1 2 3 4 5 6 7 8 9   DOC/DOC   0 9 8 7 6 5 4 3

ISBN 0-07-141053-8

The sponsoring editor for this book was Shelley Ingram Carr, the editing supervisor was Daina Penikas, and the production supervisor was Sherri Souffrance. It was set in the Gen1AV1 design in Garamond by Kim Sheran and Vicki Hunt of McGraw-Hill Professional's composition unit, Hightstown, N.J.

Printed and bound by RR Donnelley.

McGraw-Hill books are available at special quantity discounts to use as premiums and sales promotions, or for use in corporate training programs. For more information, please write to the Director of Special Sales, McGraw-Hill Professional, Two Penn Plaza, New York, NY 10121-2298. Or contact your local bookstore.

This book is printed on recycled, acid-free paper containing a minimum of 50% recycled, de-inked fiber.

Contents

2 Mountain weather   *59*

3 Reading mountain weather   *73*

4 Mountain Flying Strategy   *141*

7 Only in the mountains   *291*

8 More advanced mountain flying   *307*

Preface

Flying is a dream. All humans share the dream. Some of us are lucky enough to live it. The sky is right there all around us, easily within reach of our fingertips. But something truly amazing happens every time you take off. Three feet off the ground, and you are in another world. Don't listen to people who tell you their last flight was routine. It wasn't. It never is. It can't possibly be. Every flight is different. Ideally, every flight is wonderful. Let me think with fondness way, way back to the last routine flight I had; way back into the golden mists of time, all the way back to…the day before yesterday.

Everything in the plane was functioning perfectly. We departed less than an hour before, precisely one-half hour before official sunrise, climbing up into an already bright sky above the still dark mountains. We were already well on our way as a blinding disk of the sun broke the horizon and the highest summits, still snow-covered even this late in spring, became tinged with fuchsia far more beautiful than any description. But now the mountains were behind us. We were drifting above a seemingly flat earth spread out below us as a carpet, or perhaps more accurately, as a map. Section roads neatly divided the still sleeping world into a perfect grid of 1-mi squares. A few wisps of cloud served mainly to let us know we were actually in motion. Darker-colored ground up ahead resolved itself into the buildings of a major city. My wife and children were still asleep at home, or possibly just getting up and stoking the fire I lit before I left for the airport. Center en route traffic control handed us off to the approach controller with a more cheerful than normal "Good day." We speak to this controller at least twice a week, but we have never met him and he has never met us. He knows us only by our call sign, and we know him only as "Center," which is the form of address always used because he is the voice of the en route instrument traffic control center.

But there is seldom much traffic this early, so his tone of voice told us he remembered us from previous mornings.

I was still tired, but it was that kind of pleasant lingering tiredness that follows a good rest, rather than the fatigue I might feel at the end of a long day. I began running through the prelanding checks in much the same way that a Buddhist monk chants his mantra. And in much the same way, the mantra of the checklists fills me with an all-enveloping sense of peace and completeness.

If I complain about the long hours or the daily grind of my job, this is what I am complaining about. I fly small aircraft in the mountains for a living, day in, day out, year-round. It's time for a little honesty: My life isn't that tough.

I selectively remember most of my days like this one, but mountain flying is different. Let me reminisce again, this time all the way back to yesterday afternoon:

This aircraft is clawing for altitude well above its theoretical maximum altitude, yet the earth is no longer a map spread out below. The higher peaks tower above us to 14,400 ft. Look down. There is nowhere acceptable to land in sight. The valleys are thousands of feet deep; the ridges reach up toward you in multifingered granite spires. Even if the wind isn't blowing (and it always is), the mountains themselves are generating updrafts and downdrafts that exceed this small aircraft's ability to climb. The regional weather may be insignificant, but the mountains themselves are producing clouds out of clear blue skies and even producing their own local snow showers over the higher terrain. Altitude has robbed the aircraft of better than one-half its sea-level performance, forcing us to not just avoid, but actually exploit strange mountain weather that no one can accurately predict. Turbulence that wasn't supposed to occur is feeling very real. My seat belt is cinched down as tight as I can get it. Snowstorms that are not supposed to be there are obscuring the airport. The warm, comfortable inner glow of the previous flight isn't happening. I am offering reassurances to my passengers in what I hope is a calm, confident tone of voice. Normally it works—they have no way of knowing if this situation is normal or dangerous, so they are willing to accept my explanation.

If you apply yourself to acquiring the necessary knowledge and skills, and if you take appropriate precautions, then this flight has wonderful aspects, too, and is very acceptably safe.

On yet another not too unusual day, today in fact, I went out to the airport early to get ready to fly. Visibility was poor, but not impossible. The wind was strong, but I have certainly flown in stronger. Predicted weather over my route of flight was certainly far from perfect, but that is typical in springtime, and I have made many uneventful flights in far worse predicted weather than this. But today something told me intuitively that is would not be a good day to risk a flight. What exact combination of information sent my thinking that way is hard to say. Call it simply intuition if you will.

I don't like to cancel flights. I feel a professional responsibility to deliver when I schedule to do something. I also feel it economically. If I don't fly, I don't get paid. But today something is telling me it is best to cancel. And now that I have called off the flight, I am sitting at home, working on this book, and the weather doesn't look so bad after all. Did I made the right decision? I don't know. Have I lost a client because there are other pilots willing to fly today when I have decided not to? Again, I don't know, and I can't afford to lose clients. But at least I am on the ground asking myself that question, rather than asking it in the air. Making that kind of decision is a far more important part of my job than just flying the plane. In its own way, that too can be wonderful.

Mountain flying is very different. If you want it to be wonderful and rewarding instead of just dangerous and frightening, you need to prepare yourself.

Acknowledgments

Sources of my information

There is no magic way to suddenly acquire knowledge on this subject. It took the rest of humankind several lifetimes of sometimes very dangerous experimentation to work it all out. I didn't do that. You don't have to either. The easiest and most effective way to gain knowledge of mountain flight is to ask questions of experienced pilots. That is what I did and continue to do. Virtually every pilot I work with knows something about this subject that I don't. That is why I continue to ask questions and listen to their answers. This book is what they told me.

In short, *I did not acquire my information from out of thin air. I learned it from other people.*

These are the people who taught me most of what I know:

At Aspen Paragliding I learned to fly from Dick Jackson and Jan Stenstadvold. I later worked daily with Kelly Davis, Chuck Smith, "Dangerous Mike" Hanrahan, Laurel Pace, Rick Harris, Josh Cohen, Rauol Willie, Dave Bridges, Kirk Baker, Alejandro Palmaz, Ettienne Pinar, Granger Banks, Greg Smith, Chris Santacroce, and most frequently with Chris Blachely and David Frank. I wish to particularly credit David Frank with driving home the lesson that one good option is not enough. You need at least two alternate plans before you attempt anything in the mountains. My transition to passenger-carrying tandem paragliding only happened because of numerous flights with a very patient, cooperative, and courageous Paulina VanderNoordha. Paragliding relies far more heavily than aircraft flying on a sound understanding of small-scale weather, and these are the people who explained that to me.

Also in Aspen, Jim Shaw, chief pilot of Aspen Aviation (and a hang glider pilot), and ATA Boeing 777 captain and paraglider pilot, Shad Heaston spent many hours explaining meteorology.

At Parafly in Vermont and Mexico I worked with Rick Sharpe.

I learned to fly powered aircraft and was signed off for instrument and commercial ratings by Jim Goad at Glenwood Aviation. Jim ranks as the best instructor teaching anyone anything I have ever dealt with. He convinced me that I had the potential to fly and instruct, and he provided me with the opportunity to do both by trading flying for my work at the airport desk. His instruction gave me a framework into which all the skills and information I acquired elsewhere could fit. Jim hired me as an instructor the same day I passed my check ride. I later worked there daily with Andrea Czachor, Eric Anderson, and Joann Dunn.

Whenever possible, I would absorb anything ever spoken by Alaskan bush pilot Drake Olson, who in those days spent his winters in Colorado. In February 2002 at a backcountry dirt runway in Utah, I learned that Drake had been killed in a crash. It was 3 days before I could get anywhere within cell phone range to contact Drake's best friend, Paul Swanstrom, another bush pilot, and get details. I spent my time as people always do in those situations, wishing I had spoken to Drake one last time and told him how much I appreciated so many things he had done in his life. When I finally returned to civilization, the most important detail of the crash I learned was that it was not Drake at all. Well, Drake, in return for putting me through that anguish, your name is going in this book!

Why should I worry about a pilot with Drake's experience and judgment? The actual victim had more experience than and at least equal judgment as Drake and me combined. You can't control the outcome of every situation.

I received aerobatic instruction and guidance in becoming an instructor from Ernest Butch Thompson of Grand Valley Aviation and Hyannis Air Service; and considerable guidance in instrument instruction from Dane Koski at Arizona Aviation. My son-in-law Sean Maloney now flies DC-8s for Airborne Express and is a former Cape Air and Allegheny Dash 8 captain. He has spent many hours guiding me through the intricacies of the aviation business.

I was introduced to gliders by "Glider Bob" Saunders, who telephoned me with a job offer the day I passed my commercial check ride, acting on a recommendation from the FAA examiner. In checking my logbook, I see that I towed up "Glider Bob" nearly 600 times in one very busy summer. I was taught to fly gliders by LaVerne St. Clair of Durango Soaring.

At Telluride Soaring I worked regularly with Vincent Esposito, John McKillop, Norm Grey, Nick Kennedy, Todd Lipke, Jim Lincoln, Dan Schrank, and Gavin Wills.

I am particularly indebted to glider instructor and airline pilot Brian Lewis for the use of some outstanding photographs of mountain weather reproduced in this book. Commercial helicopter pilot Laurent Giles worked on the illustrations. Pilot/photographers John and George Kounis are the editor/publishers of *Pilot Getaways Magazine*. Pilot/photographer Gerrit Paulson is editor of *Southwest Aviator Magazine*. Ron Kanter of Telluride Aviation Air Photography is a former C141 Air Force pilot as well as a flight instructor and glider instructor.

At Mountain Aviation Services I work daily with company dispatcher (and my wife) Shelby Evans. We have just arranged to work more regularly with Lighthawk check airman and former Aspen Flight School director and chief charter pilot Dick Arnold. Dick very courteously reviewed a draft of this book and filled the pages with so many sticky notes that I spent two 10-h days doing revisions. He spent a couple of days making other, very valid suggestions, sending me to the computer all over again. I don't know if thanks are really in order to any one forcing me to do that much work, but thanks are certainly in order for the information. I hasten to add that errors which remain are mine, not Dick's.

The recommendation which got me my first job as a corporate pilot came from charter pilot, transatlantic ferry pilot, Hollywood stunt pilot, test pilot, and aviation legend Peter Lert. In addition to a lifetime of professional flying, Peter was subjected to a lengthy classical education and has a second skill and career as a writer of at least equal merit, if not fame. Once I had struggled past Dick Arnold's criticisms, I was subjected to an even heavier dose from Peter. I don't need enemies to beat me up. I have friends who can do an even more severe job of it! Honest, sincere, constructive criticism is something only real friends can provide. Thanks, even if I was really abused badly. Again, errors which remain are mine, not Peter's.

I have traded as much information as possible with Jeff Wood of Black Canyon Aviation and Superior Air Freight. I am not at all embarrassed to grovel at the feet of and pick the brains of Blake Freeland of Cimmaron Aviation, Larry Kempton of Kempton Aviation, Michael Gardiner of Flight One, Bob Wallick of Falcon Flight; charter pilots Terry Sargent, Todd Wilson, Josh Thompson, Eliot Brown, and Galen Rassmussen; and airport manager Rich Nuttall of the Telluride Regional Airport. I am particularly grateful to Jeff Garcia and Troy Beattie, both formerly senior pilots and company check airmen for Great Lakes Airlines, who provided invaluable information on flying back and forth to Denver in instrument icing conditions. I have been Grant McCargo's copilot on hundreds of scheduled flights over the continental divide between Telluride and Denver.

Whenever they land, I make a point of rushing forward to ask local pilots such as David Wright, Eric Cohen, Dale Wood, Jim Lucarelli, Steve Hilbert, Dean Gianpietro, and many others what they have encountered. The counsel of these and other skilled pilots whose judgment I trust implicitly can have a profound influence on my safety and that of my passengers.

I have tried to read anything ever written about mountain flight. The best material is of course that written by Ernest K. Gann, particularly in *Fate Is the Hunter,* and Tom Wolfe's *The Right Stuff.* The most useful, though, is by Sue Baker and Margaret Lamb; I intend to attend any seminar they ever offer, and I advise everyone else flying in the mountains to do the same. Sparky Imesson has rewritten his treatise *The Mountain Flying Bible* several times, and I recommend you purchase each subsequent edition. It covers some of the same material as this book, but with more technical details and greater emphasis on larger aircraft and instrument flight. My intention is never to actually use the information in Mick Wilson's *How to Crash an Airplane,* but if the occasion arises, I am sure I will be glad to have read the book before the fact. Many of the statistics in this book derive from a variety of primary sources, in particular, from the U.S. Department of Transportation, but I knew where to look for those primary sources as a result of encountering similar material in the writings listed here.

FAA designated examiner Lynn French has made every check ride a lesson rather than an examination and suggested uncountable ways to improve my teaching. (And in his other capacity as an attorney and a judge, Lynn presided at my marriage ceremony.) I struggled through the Air Carrier Certification process with help from Albert Rohr of Flight One Aircraft Management and Ron Livingston of Aircraft Business Consultants. Our charter certificate is administered by FAA representatives Dave Cawthra and Bob Lesitsky at the Salt Lake City Flight Standards District Office. It has been tremendously helpful of them to provide guidance as well as regulation, guidance which has found its way into these pages.

After reviewing the Mountain Flying and High Altitude Operations standardized training program developed by Bruce Hulley and Vern Foster for the Colorado Pilots Association during the preparation of this book, I was recruited into the program as an instructor by Jere Eberhardt.

I had the exceptionally good fortune to find myself working around these particular people, and therefore I find myself in a position to listen to their advice. You will discover that wherever you are based or flying, there are plenty of very experienced and very helpful pilots around, and you would be well advised to ask them questions, just as I did.

The nature of teaching is that you learn as much as or more from your students than they learn from you, and I hope I paid enough attention.

Here and there, this text examines examples of poor flying judgment. For the most part, these involve honest mistakes by good people who were humbled and who learned not to make the same mistake again. There are also some very bad, even suicidally stupid examples of mountain flight and instruction to be found in this book. Virtually all these bad examples are provided by commercial pilots. That selection is intentional. Not just neophytes make errors in judgment. We all routinely make mistakes, and in retrospect we all do stupid things that didn't seem like such a bad idea at the time. With the perfect benefit of hindsight, it is always very easy to tell what someone else did wrong. Most of us, however much we might wish otherwise, learn by making mistakes. We survive in part because it usually requires a chain of mistakes to get you so deeply in trouble that you can't escape.

The names of pilots providing us with both honest mistakes and glaring examples of stupidity are left out for those reasons, as well as a desire not to brand them forever with incidents from a past that they, as well as the rest of us, may have learned from.

This list is an incomplete one and I intend no slight to those whose names I have left out. This list is also shameless name dropping, and I apologize to those named only for absorbing so incompletely the information they have to offer.

A word of caution

This text is written as an aid to actual flight experience and instruction in the mountains, not as a substitute for it.

Like any book, it is subject to both interpretation and misinterpretation. A living, talking human instructor experienced in mountain flying right there with you in the airplane is a better source of instruction by far. You can ask the actual human being questions and clarify things. He or she can translate theory into practice. The instructor can identify a potentially dangerous situation before it gets out of hand, explain it to you right there on the spot, and help you get yourself out of it.

You can find a flight instructor with mountain flying experience at virtually any mountain airport. You can find one at most flight schools located in major cities along the mountain front, such as Denver or Salt Lake City. Indeed, although those two cities are well outside the mountains proper, some of the most demanding mountain flying you may ever encounter can occur while climbing out from either city and trying to cross the nearby mountains. Before you actually fly into the mountains, don't just read about it—get some actual flight training with an instructor who specializes in the subject.

I would like to recommend, as a first step, the Colorado Pilots Association Mountain Flying course. This course, which I am scheduled to help teach, provides standardized ground school and mountain flight experience with a qualified instructor over standardized routes through the mountains. The program was created by Vern Foster and Bruce Hulley in 1986. It has been refined ever since and has introduced more than 2700 students to mountain flight through its one-day ground school, and it has flown 500 of them in the course. The Colorado Pilots Association has trained students from Iowa, Nebraska, Minnesota, Illinois, Kansas, Oklahoma, Texas, California, New York, Missouri, New

Mexico, Indiana, Ohio, Mississippi, England, Germany, and of course Colorado. See www.coloradopilotassociaton.com.

The instructors involved with this program teach independently as well, and a list of their names is available from the Colorado Pilots Association.

Introduction

Mountain flying is different

There are as many reasons to fly in the mountains as there are pilots. Scenery found nowhere else in the world would be justification enough. To that you can add the challenges and rewards of attempting to fly harmoniously with the forces of nature, and the mundane needs of practical basic transportation in a part of the world where roads are few and indirect. Make up your own excuses. The fact is, mountain flight is tremendously exciting and tremendously rewarding.

Mountain flying can be done safely

Flying in the mountains in a small aircraft is a different experience from flying in any other environment. Even on the very best of days, it involves considerably more than just normal flying over exceptionally scenic terrain. The aircraft's engine develops only a fraction of its rated horsepower at high altitude. Because the air is thinner at altitude, a typical small plane on the ground at Telluride at over 9000 ft above sea level develops only 65 percent of the horsepower claimed in the operator's manual. Because the air is thinner, the wing needs greater true airspeed to develop adequate lift. Here in Telluride, it needs 20 percent higher true airspeed. The pilot may also suffer from lack of available oxygen at altitudes lower than required to clear the mountains. Steep and high terrain is hard to fly over. Navigation is frequently indirect and has to be done without full use of navigational aids. Weather conditions which exceed the operating capabilities of most small aircraft are frequent, normal occurrences, and weather can change dramatically in minutes.

It is possible to fly very safely in the mountains. It requires no exceptional ability or secret knowledge, but it does require training, knowledge, and experience!

The average pilot can learn to deal with all the aspects of mountain flight and fly with a very acceptable safety margin, but it doesn't happen by itself. You have to study and work at it. You might be able to figure it all out by yourself through trial and error, but it will be faster, easier, and safer to benefit from the combined experiences of other mountain pilots.

The statistics of mountain flight

Flying in the mountains is statistically much more hazardous than flying in flat country.

Despite the obvious hazards, aviation has proved to be the safest form of transportation ever invented. But flying a small airplane in the mountains is by no means inherently safe. Pilots tend to dismiss some of the hazards of flying by comparing it with the much higher accident rate for automobiles. This comparison is not a valid one for small aircraft in the mountains.

The statistics should scare any pilot venturing into the mountains for the first time. Mountain flight in small aircraft is much more dangerous than driving.

How much more dangerous? In 1989 Susan Baker, an epidemiologist, risk management specialist and medical researcher whose credits are too lengthy to list in a single book, and Margaret Lamb, a charter pilot, fight instructor, and weather expert, teamed up to study aircraft accidents in the Aspen, Colorado, area for the period from 1964 to 1987. Before the study began, it was generally conceded that the accident rate in that area was high. During the study period, no less than 232 airplanes crashed within a 50-mi radius of Aspen, resulting in 202 deaths and 69 serious injuries.

The General Accounting Office has concluded aircraft are 40 percent more likely to crash in the mountains than anywhere else. According to the National Transportation Safety Board, the aviation death rate in mountain states is twice that of the nation as a whole. Although other statistics are less reliable, it should be intuitively obvious that the accident rate and the danger in the mountains are actually higher still because there are fewer people actually flying in the mountains.

Year after year, the greatest number of aviation accidents in the United States occur in Alaska, Washington, Florida, Texas, and Colorado. Drop Florida and Texas from that list. Their higher accident numbers are the direct result of very high populations and lots of aircraft flying in crowded airspace. In addition to mountains, Alaska has the obvious extremes of vast distances and poor weather from the nearby oceans, and no roads, so everyone flies. Most of the necessities of life travel by small plane. Some very skilled, very experienced Alaskan pilots find themselves pushing the limits because what they are carrying is very important to someone. Try someday to get commercial insurance for a small plane in Alaska—see what the insurance company thinks about the risk!

So take Colorado as our example. As a mountain state, it exceeds the very high aviation accident and fatality rates of all but two other mountain states, yet most of Colorado is not mountainous. And 90 percent of Colorado's population lives in the corridor of major cities east of the Front Range, out of the mountains. Most of the state's flying traffic is there. West of Denver, the rest of the population is concentrated out beyond the western edge of the mountains proper, and most of the flying in western Colorado takes place west of the mountains. Yet the statistically largest part of Colorado's flying accidents occur in the mountainous central third of the state where only one-tenth of the state's population lives and much, much less flying is taking place.

If my goal in writing this down were to scare you out of even considering mountain flying in a small aircraft, I could stop writing now. You should be scared. These are *very* significant numbers.

Obviously that is not my goal. (After all, here I go with another paragraph.)

I don't want to scare people away. I do want to help them fly safely. I derive my livelihood from flying small aircraft full-time, year-round in the Rocky Mountains. I fly with hundreds of people every year. I have a wife and children, and I live a fairly quiet, normal life. I am not the sort of person to take extreme risks on a regular basis. Rather, like most pilots based in the mountains, I have been taught to recognize and avoid many of the unusual and specific risks of mountain flying.

Am I deluding myself? Or is there in fact a safer way to fly the mountains? I believe there is a safer way, and to a degree statistics back me up.

Causes of mountain flight accidents

Causes of crashes in the mountains are different from causes of flatland crashes.

In flat country, the leading cause of small aircraft accidents has long been unintended visual flight into instrument conditions. That type of accident is less common in the mountains. These types of accidents generally fall into a category referred to in bureaucratese as "controlled flight into terrain." Only 24 of the 232 crashes in the Aspen study fell into that category.

According to the Colorado Department of Transportation and the Colorado Division of Aeronautics, the biggest category of crashes in Colorado (26 percent) includes as a cause "loss of control in flight"— a rare cause of accidents elsewhere, and a testimony to the degree of turbulence and up- and downdrafts you can encounter in mountainous terrain. To that add "in-flight encounter with weather" (20 percent) and "high density altitude" (100 percent), factor in an inability to outclimb terrain (65 percent), and a pattern emerges. The leading cause of accidents in the mountains is the failure of pilots to take into consideration the effects that the mountains themselves have on wind, weather, and aircraft performance.

Again, the flying world owes a considerable debt to Margaret Lamb and Susan Baker. After the initial statistical analysis, Lamb and Baker strapped on crash helmets and flew Lamb's Navion through the scenes of many of the accidents, often in very similar kinds of weather. This can only be described as a very gutsy move. After all, others had crashed in similar circumstances, and the crash helmets were necessary to protect their heads when they bounced off the sides of the canopy in turbulence. In the course of this process, they discovered a variety of then poorly understood weather phenomena and later conducted numerous seminars to share that information with other pilots.

It is noteworthy that various aspects of the mountain environment were significant contributing factors in most mountain accidents, and contribute to some degree in virtually all mountain accidents, including even those attributable to mechanical failure. Were those factors not present, most of the crashes would not have occurred, despite other contributing causes.

To eliminate all the hazardous factors of mountain flight and the associated risks is impossible. Understanding those factors provides a basis for taking actions to reduce the risk. Once these special mountain factors are understood and allowed for, the safety of mountain flight approaches, but never quite reaches, that of other kinds of flying.

Who has mountain flying accidents?

Again, I refer to the Baker and Lamb 1989 study. Pilots crashing in the Aspen area were by no means all low-time pilots, although pilots with less than 100 h *in the type of aircraft they flew* were seriously overrepresented, accounting for 44 percent of crashes. *Lower-powered, four-seat aircraft* were overrepresented, but this must be seen in the context of the fact that the majority of aircraft flying everywhere in the country fit into this category. In part coincidentally and in part for the same reason, general aviation was very overrepresented. This is in contrast with flying accidents elsewhere in the world. The Cessna 172, the most produced and flown aircraft of any type in history, is the definitive low-powered, four-seat aircraft. Worldwide, it has the best safety record of any aircraft ever made. Whether it can realize that same safety record in the mountains is up to the pilot.

Many experienced pilots will simply tell you not to fly a low-powered small aircraft in the mountains. While I don't disagree with that advice, I nevertheless fly that kind of aircraft all the time, and so do thousands of other mountain pilots. This text is *intended* for pilots of that very type of aircraft.

In the mountains, men crash more often than women.

Why? Nationwide, the leading cause of flying accidents for women is loss of control of the aircraft. The leading cause of accidents for men is failure to exercise good judgment. In the mountains, exercising good judgment is more important than good flying skills. As the late Butch Thompson so often stressed: A superior pilot is one who exercises superior judgment so that he or she doesn't have to employ his or her superior skills. Machismo is not an asset in the mountains.

Men and women have slightly different emotional makeups: Most men, including me, psychologically feel a sense of personal defeat if they have to ask advice. Women, on the other hand, feel empowered by gaining more knowledge.

Take my advice—fly like a girl!

Instructional flying was overrepresented, and at least anecdotally it would appear that instructors from outside the mountains were part of the reason. A lot of mountain flight instruction is being given by very low-time flight instructors based outside the mountains whose own mountain experience is limited to a couple of "mountain checkouts" which were in turn given by other instructors with a similar lack of mountain experience. A couple of brief forays into the mountains in optimal conditions is definitely very valuable and an excellent starting point for mountain flying. It is not enough to qualify someone to teach other people mountain flying. A day of ground school and a 4-h training flight from Denver up into the mountains on a perfect day as part of the Colorado Pilots Association Mountain Course is tremendously beneficial, and pilots who take a course like that one become much safer, but hardly qualify yet as mountain experts.

Are you an instructor planning to teach mountain flying? There are two ways to be much better at it. One is to attend seminars. The Colorado Pilots Association has sponsored seminars specifically for mountain flight instructors. The other thing is to serve an apprenticeship with someone skilled in mountain flight. Most large flight schools in or near the mountains will have a very qualified chief instructor.

Newly minted flight instructors, commercial pilots, military pilots, and doctors tend to be very apt students, but not automatically safe ones. They are typically at the very peak of their learning potential, they are of above-average intelligence, and they are highly motivated. These people are an instructor's dream. They are very easy to teach. They learn unbelievably quickly. Why, then, do they have more, rather than fewer, mountain accidents than some other pilots? Are they simply careless risk takers? Most of them would take considerable offense at the suggestion. They regard themselves as highly professional, and I agree.

The risk for very motivated, very fast learners is that they tend to be overconfident and impatient. They tend to minimize the importance of preliminary training and basic skills, focusing instead on their final objective without giving consideration to the means employed to get there. Because they make very few mistakes in training, they tend to assume that the correction of the few errors they do make is not terribly important. If you think of this in terms of a written test, the student who scores 98 percent gets an A. That student tends not to worry very much about the one or two questions missed. The student who scored only 70 percent nearly failed and is very concerned about the other 30 percent.

Mountain flight is only a very small portion of most pilots' flight training. If they got the other 98 percent right with ease, how important can that last 2 percent be? It is not important at all if you manage to fly only on perfect days. It is being too casual about the tiny portion of their flying that will actually be in the mountains which gets these very capable, very qualified pilots in trouble.

Flatland pilots are at greatest risk

One factor leaps from the statistics more than any other: Baker and Lamb's statistical analysis revealed that no less than *41 percent of the pilots crashing in the study area were out-of-state residents.* Something like only 10 percent or possibly fewer pilots had 41 percent of the accidents.

Outside the study area in the rest of Colorado—an area which still includes plenty of mountains—out-of-state pilots accounted for only 25 percent of all accidents. Let us apply a bit of conjecture to that number. Of Colorado residents, 90 percent *don't live in the mountains.* Most Colorado pilots don't live in the mountains and do most of their flying out of the mountains, occasionally venturing into the mountains only on perfect days. To those 41 percent flatland out-of-state pilots should be added an unknowable but probably very large number of flatland-based, -trained, and -experienced Colorado pilots.

I don't mean to imply that mountain experience makes you a better or safer pilot. It doesn't. It only makes you safer in one specific way. Crushing as it may be to our egos to hear it, outside of the mountains, mountain pilots are no more skilled, no more knowledgeable, and no safer than any other pilots. Yet most of the flying in the mountains is done by mountain-based or mountain-experienced pilots, while a significant number of accidents there involve occasional visiting pilots from outside the mountains. This is not to say that mountain experience confers some form of invulnerability on pilots. Far from it. *Mountain-trained and -experienced high-time pilots in familiar aircraft still have accidents in the mountains.* The causes tend to be similar to those of everyone else's accidents, and those causes often relate directly to the mountain environment. It's just that pilots with appropriate experience and training have these particular accidents less frequently.

In the mountains, mountain pilots are substantially safer than other pilots. In instrument conditions, instrument pilots are safer than VFR pilots. In class B airspace, urban pilots are safer than rural pilots. There is no magic here.

Most mountain pilots don't realize that they have been the beneficiaries of some very specialized instruction and experience. After all, if you have never taken off from an airport lower than 5000 ft mean sea level (msl), then you just assume that all aircraft are very low-performance and fly accordingly. If downdrafts exceed your ability to climb every day that you fly, then you are always expecting them and are always prepared to take appropriate action. Deliberately or accidentally, mountain-based flight instructors teach these skills every time they go out. Mountain pilots practice them every flight.

The lower accident rate of mountain experienced pilots in the mountains indicates something else even more important. **Safe mountain flying skills, planning, and techniques are generally understood by most mountain pilots and can be learned by anyone.**

If this is true, then it is clear that mountain-trained and -based pilots share a wealth of experience, skills, knowledge, and attitude which somehow allow them to fly in the much more hazardous environment of the mountains with an accident rate not terribly worse than flatland pilots enjoy everywhere else in the world.

The risks of mountain flying are inherent in the mountain environment and cannot be eliminated. They can be mitigated by using certain specific skills and acquiring certain mountain-specific knowledge.

Hence this book

I am writing this text for the usual motives of profit and vanity, but I am writing as well in the firm belief that I can make many pilots' flying much safer.

I live 2 mi from the Telluride, Colorado, Regional Airport—at 9080 ft msl, the highest commercial airport in the United States. For me this is a big step up in life; in fact it is a step of 3164 ft. The last place I worked, I lived 200 ft from the runway centerline of the Glenwood Springs, Colorado, Municipal Airport, at only 5916 ft msl. Although Glenwood Springs is often called the toughest mountain airport in

Colorado, I flew my first powered aircraft solo flight there. It was no problem; I had no special talents, but I was well taught. I have soloed dozens of student pilots at both Glenwood and Telluride. I am only an exceptional instructor by my own self-serving standards. Only some of my students are gifted. What they all share is experience in the environment in which they are flying.

I work here every day. Although I do some corporate flying, charter, air photography, game spotting, environmental flights for Lighthawk, gliders and glider towing, and occasional search and rescue, much of the daily grind (if it could be called that) of my job involves introducing visiting pilots to mountain flying.

Because I fly on a schedule, and because a lot of the flying I do involves being very close to inhospitable terrain, a normal instinct for self-preservation has made me take a particular interest in safe mountain flight. A concern for my students has led me to try to formalize some of that knowledge.

I know most of the local pilots and meet many of the visiting ones. That means that often when there is an accident, it involves someone other than a perfect stranger. For me, as for most pilots working in the mountains, there is often some connection with the victims and with it a sense of personal loss. Because I am one of the very few commercial pilots at our airport working primarily with small aircraft, and because I am the only search and rescue pilot for the local sheriff, I often find myself involved in the aftermath of the accident or the search that follows.

I do not claim to be the guru of mountain flying. At best, I am an avid student of the subject. What knowledge I have on the subject is shared by the overwhelming majority of professional mountain pilots and instructors. Most of what I know about flying I learned by the incredible good fortune of meeting and being guided by people who know a great deal more about the subject than I do. The only difference between them and me is that while they are out flying, I am taking the time to organize and write some of this down.

Nothing in this book is original research. I am not inventing any new concepts. I am just sifting through and organizing the things other people have taught me. This is a compendium of the basic skills and information that all experienced mountain pilots use every day. This is the how and why of their basic safety procedures.

The knowledge we share is not well known or well understood by pilots flying outside the mountains.

The March 2002 issue of AOPA *Pilot* magazine contains an article by Crista Worthy entitled "Caught by the Wave" about an extremely dangerous flight with *two* encounters with what the author describes as mountain wave over the Sierras, but what this book would call rotor turbulence. In 72-knot winds at 13,500 ft the *25-year veteran Navy pilot* flew over a sharp ridge only 500 ft off the ground. The next 8 min of flight was life-threatening extreme turbulence. The attempted flight home was more of the same, and he prudently turned back.

I fly the same type of aircraft over mountains 2000 ft higher in stronger winds every week, and believe me, if my flights were as dramatic as that, I would probably not try it again! Even if I would, no one would ever fly with me. I fly the same type of aircraft as in the magazine article in the same or worse conditions, but I do it slightly differently. This book explains how it is done. I am not prone to take big risks. I pretty much just grind back and forth every day over the mountains, just making a living. Okay, maybe it is not quite a routine, boring daily grind. What I do is absolutely exhilarating. The scenery alone is mind-boggling, no matter how many times I see it. The skill of finessing a small aircraft through the overpowering forces of nature shares the same appeal as sailing, skiing, or any number of other sports. The reality is that I must derive something special from it since I used to be paid considerably more to sit at a desk and work for someone else. But because I basically understand what I am doing, it never rises to the degree of excitement described in the AOPA magazine article. For example, in that much wind, you would either find me at least 2000 ft higher or see me turning back very early.

Yet every year, life-threatening experiences like those in the article do happen to the majority of new pilots visiting the mountains, and a small number actually crash because of their lack of knowledge.

The most fundamental safety considerations of mountain flight are a complete mystery to flatland pilots. A professional pilot with 25 years' experience did not know the basic information people like me have to rely on every day of the week. None of us would have crossed the same ridge in the same conditions without considerably more altitude. Does that mean he was a bad pilot? Not at all. Do you want to see a bad pilot? Watch me try a carrier landing with no more aircraft carrier experience or instruction than he had mountain instruction. Can I fly better in the mountains than he can? Probably not. I just choose to fly different routes in those conditions.

Reading could help if you found the right book. There are books and articles on this subject, some of them excellent, but those books are not generally read by all the people who could benefit from them. I list a number of very useful books in Chapter 8 under "Further Reading." Some of them are slow-going; most are simply hard to find.

I hope that I can put this subject in a simpler, shorter, more readable form. I am trying to resist every author's temptation to impress you with all the extraneous information I command and selectively cull the basic information I think you need. If I can do that, then more people will be able to access and actually use the information. That in turn will make those pilots' mountain flying what it should be—enjoyable, exhilarating, practical, and safe.

Disclaimer

No text, including this one, is a substitute for actual instruction. You can learn safe mountain flying by a gradual, cautious program of trial and error with no help at all. You would certainly do much better if you were guided by this book. But the very best way to learn mountain flying, and certainly the safest, is to fly with an experienced mountain instructor. This text is as accurate as I can make it, but it inevitably contains errors and omissions. No textbook can guarantee your safety.

I don't know everything, and when I learn new things, I am still capable of changing my mind. At some time in the future, I might possibly want to substantially revise parts of this book. You can make the same revisions I would. How? Don't be afraid to ask for a second opinion. If you know other pilots with mountain experience, ask them questions. If they disagree with passages in this book, ask why and consider both opinions.

1

Altitude

Mountain flying is tremendously affected by altitude.

Effects of altitude

Some traces of the earth's atmosphere may be detected at altitudes as high as 350 mi. By convention, though, space is said to begin at 100-mi elevation, the realm above which all flight is ballistic rather than aerodynamic. The lights of the Aurora Borealis result from the interaction of solar radiation and the atmosphere at elevations from 70 to 100 mi high. The very highest of cirrus clouds are at altitudes of only around 7 mi, which we can now convert to a more useful 35,000 ft.

Yes, some of the atmosphere reaches up very far away—but not very much of it. The atmosphere in which conventional aerodynamically controlled aircraft can fly is very thin indeed. Few airliners can reach 40,000 ft—less than 8 mi above sea level. Yet this is a very deceptive number indeed, because *in terms of pressure, one-half of the earth's atmosphere is below 18,000 ft. Most of the earth's weather is also below 18,000 ft.* One-half the atmosphere, one-half the weather, nearly all the clouds you can see other than large thunderstorms—all that is less than 4 mi higher than sea level.

At elevations over 14,000 ft, the Rocky Mountains rise almost 3 mi above sea level. You reach halfway through the atmosphere only 1 mi higher. The atmosphere is a very thin shell around the earth, and high mountains poke almost halfway through it.

Any thought of flying in the mountains must be put in the context of altitude. This book has as its basic frame of reference the Rocky Mountains of Colorado, where no less than 52 mountains exceed

1

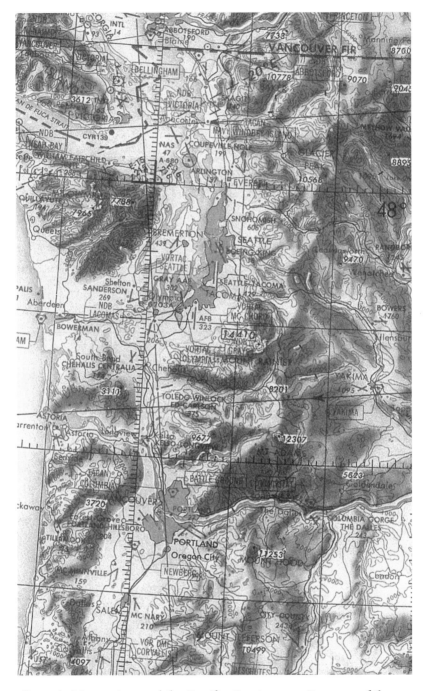

Cascade Mountains and the Pacific Coast range. Because of the nearby ocean, these are the rainiest and snowiest mountains in America.

14,000 ft in elevation. But elevation is a factor to varying degrees in all mountains. The Saint Elias Mountains in southeast Alaska rise to even higher elevations right out of the ocean. The less than 5000-ft elevations above sea level of all but a couple of the Appalachian Mountains in the eastern United States might seem trivial compared to the climb from over 5000 ft up to over 15,000 ft needed to clear the Front Range only 20 mi west of some Denver airports. Yet that mild rise in elevation compared to the substantially lower Mississippi valley to the west and Atlantic coastal plain to the east can and frequently does produce severe weather which exceeds the capability of small aircraft. Gently rounded Mount Washington in New Hampshire, scarcely 5200 ft high above a 3000-ft-high base, sees surface winds over 100 mi/h. Your aircraft may have far better performance flying the low Appalachians than the high Rockies, but you are still faced with at least as much mountain-induced weather, if not more. I direct the reader's attention to Ernest Gann's *Fate Is the Hunter* (see "Further reading" in Chapter 8) for descriptions of low visibility approaches and icing over the Appalachians.

A few peaks in the Sierras are quite high, but an altitude of 12,000 ft will see you over virtually all the rest—unless of course the very nearby Pacific Ocean has added moisture to the air. Then snowfall in places such as Donner Pass can be so rapid and so intense that people are stranded for months and forced to eat their dead. Just east of the Sierras in the Sierra wave, glider altitude world records are set on good days, and very large airliners are tested to the point of airframe damage on bad ones.

Within the mountains proper in Colorado, the lowest airports are still close to if not above 6000-ft elevation, with Telluride at 9078 ft, Leadville at 9927 ft, and a large group clustered just below 8000 ft. This last figure is determined by climate—8000 ft is the lowest elevation in Colorado where you find dependable snow cover all winter long, and therefore it is the base elevation for most ski areas. Sitting on the ground on the ramp at the airport in Telluride, Colorado, you are feeling and breathing air at less than 70 percent sea-level pressure. The altitudes required to safely cross the 52 peaks of the Rocky Mountains with altitudes over 14,000 ft in marginal weather will have you flying in air at less than 50 percent sea-level pressure. Your airplane and its engine depend on air. Here, with one-half the air, they will do only one-half as well.

The northern Appalachian Mountains. Mt. Washington in northern New Hampshire has recorded a record 265 mi/h wind.

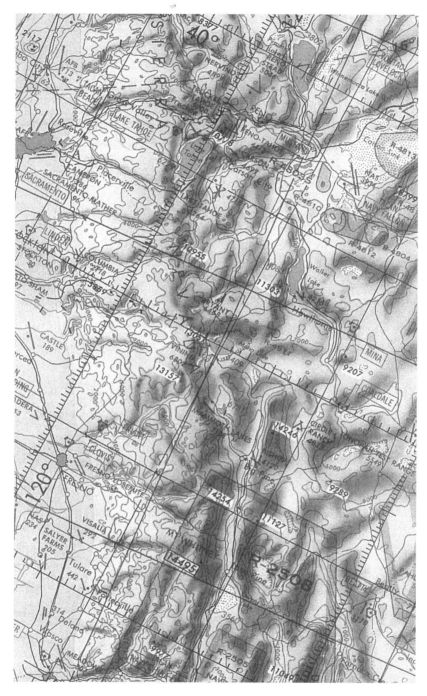

The Sierras: The Owens valley east of Fresno and Mt. Whitney. Sees mountain wave reaching over 50,000 ft msl. Travelers at Donner Pass north of Lake Tahoe were trapped by heavy snow and ate their dead.

The central Rocky Mountains. Some 54 mountains in Colorado are over 14,000 ft high.

Altitude, in the context of flying and breathing, is not so much a factor of how many feet above sea level as it is a factor of air pressure. On a hot day or a humid day, the air is much thinner, so the effective altitude can be substantially higher. More on this later....

Both your body and your aircraft require air, and there is nowhere near as much of it available as you want in the mountains.

Some aircraft performance charts

Service ceiling is the altitude above which an aircraft cannot maintain a climb rate of 100 ft/min. For all practical purposes, this is about as high as you can fly. *Absolute ceiling* is a little higher but, from a practical standpoint, very difficult to actually maintain. Let us take a look at the service ceiling of some typical low-performance small aircraft:

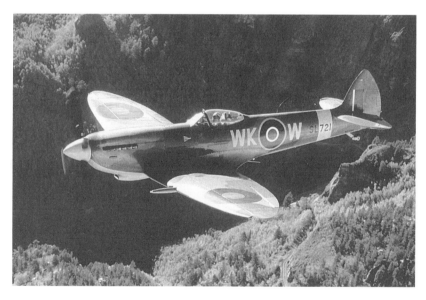

This might be the perfect mountain small aircraft. It is no bigger than a Beech Bonanza, but with 1600 hp, the Spitfire has a climb rate well in excess of 2000 ft/min and can in fact pull up into a vertical climb from horizontal flight and gain better than 1000 ft before having to level off again. Endurance is only about an hour, luggage capacity is limited to the wing storage bays where the machine gun ammunition used to go, and this one is for sale for close to $2 million—but you want one, don't you? TR Youngstron photo.

Aircraft	Service ceiling, ft
Aeronca	
15 AC Sedan	13,000
11 CC Super Chief	12,500
7 DC Champ	13,000
7 AC Champ	12,500
11 AC Chief	11,000
American Champion	
7 ECA	12,000
7 ACA	8,500
American General/Grumman	
AA5, AA5A	12,650
AA1B Tiger	12,750
Beechcraft	
C23 Sundowner	12,600
A23, A23A Musketeer	11,870
B 19 Sport	11,650
Skipper	12,900
Cessna	
Early 206 on skis	11,500
	(versus 16,700 without skis)
177 Cardinal, 150-hp	12,700
150 K, Aerobat K	12,650
	(versus 14,000 for a 150 M with same engine)
Lake	
C1 amphibian	9,500
Maule	
M4, M4C	12,000
Piper	
PA 32-260 Cherokee Six	12,800
PA 28-235 Cherokee Charger	12,000
PA 28-161 Warrior II	11,000
	(this is not a misprint)
PA 28-140 Cherokee	10,950*
	(neither is this)
PA 22-108 Colt	12,000
PA 14 Cruiser	12,500
PA 12 Super Cruiser	12,300

Aircraft	Service ceiling, ft
J3 Cub	11,500
Rockwell	
Lark-180 hp	11,000
	(and there is a 150-hp version which does even worse)
Socata	
Tobago	13,000
Tampico	11,000

*Two of the most popular four-seat aircraft in the world cannot even reach pattern altitude at Telluride, Aspen, or Leadville, Colorado, on a typical summer day!

Never mind flying over a 14,000-ft peak. Upon flying west out of Denver, Colorado, the ridge above the Eisenhower Tunnel on Interstate 70 is higher than 13,000 ft mean sea level (msl). Safe crossing altitude is over 14,000 ft msl. Not one of these airplanes could be expected to cross the 12,000-ft Loveland Pass just west of Denver with a full load on an absolutely optimal day, and the density altitude of that ridge on a summer day might be 2000 ft higher. Of the mountains in Colorado, 52 are more than 1300 ft higher than the service ceiling of the *best*-performing aircraft in this list, and 5500 ft higher than the worst-performing.

The above sample was selected to show the large numbers of general aviation aircraft which *cannot* fly as high as most, if not all, major mountain passes in Colorado. At lower elevations, these are very popular, very useful aircraft. Any of these would be a very impractical, not to mention dangerous, aircraft for a mountain airport.

In retrospect, many accidents involving aircraft on this list seem very predictable. (But then with the benefit of perfect hindsight, it is always easy to tell what someone else did wrong.) In the early summer of 2001, a heavily loaded Piper PA 28 Cherokee 160 headed east over the Flat Tops primitive area east of Meeker, Colorado, simply sunk into the ground, and crashed in a lake. As the name implies, the Flat Tops are a broad, flat-topped mesa many miles in extent with an average elevation over 12,000 ft. The aircraft's service ceiling was less than 11,000 ft. As the day got warmer and the air thinner, the plane simply could not stay airborne. Never mind the fact that the aircraft was overloaded or the pilot's toxicology suspect. The cause of the accident was extremely simple: The airplane could not fly that high. No amount of pilot skill could change that.

The following summer, another Piper PA 28 Cherokee left Eagle, Colorado, headed the other way over the same terrain, this one loaded with four adults plus a 2-year-old child in a car seat. It, too, simply sank into the ground and crashed with fatal results. Never mind that it, too, was overloaded—that make and model of aircraft simply cannot fly as high as the surface of the ground in that part of Colorado.

People reviewing this text have wondered why such aircraft are ever produced. Flown over the major part of the country outside the mountains, these aircraft almost all climb 10,000 ft higher than the ground. Flown over the 5000-ft-high Appalachians, most of them still can get better than 5000 ft higher than the highest terrain. That is far more than adequate for 90 percent of the country. It is only a problem in the other, mountainous, 10 percent.

You can manage to fly the four- and six-place aircraft on this list in the mountains by flying them as two seaters, or some of the small two-place aircraft solo. But why do you want a six-seat aircraft that you can only fly with two people and no luggage?

My neighbor and sometimes coworker Todd Wilson flies an 85-hp Cessna 140 from his over 9600-ft elevation grass runway; but once he is airborne, his rate of climb even when flying solo is very mild. In fact, he really doesn't climb at all. The runway is on top of a high mesa. He just takes off and flies along level. He has flown that particular aircraft out to the West Coast over the Sierras and east to New England over the Rockies. Still, he would be the first to admit that on certain days his Cessna 140 is best left on the ground despite its theoretical 15,600-ft service ceiling, and he instead drives his car to the Telluride airport, 6 air miles away, to go to work flying twin turboprop charter flights.

What small, basic aircraft are acceptable for mountain flight? Some better choices from the same manufacturers might be the following, to name only a very few:

Aircraft	Service ceiling, ft
Aeronca	
7CCM Champ	16,000
Champion	
7GCB Scout	18,000

Aircraft	Service ceiling, ft
Aviat	
Husky	20,000
Beechcraft	
35, 36 Bonanza (various models)	16,000–22,000
33 Debonair	18,000
Cessna	
T210R	29,000
P210	24,000
210D	21,000
TU206	24,600
206—nonturbo	16,700
182 (various years)	17,700–20,100
	(nonturbo only— turbos much higher)
172 (various models)	13,100–17,000
	(engines 145–180 hp, gross weights 2300–2550 lb)
170 A, B	15,500
152	14,700
150 M,L	14,000
140 A	15,600
120	15,500
Liberty	
Aerospace	16,000
Maule	
M6, M7, M8, 235 hp	20,000
Piper	
PA 24 Comanche 260	20,000
PA 22-150 TriPacer	16,000
PA 18-150 Super Cub	19,000

From these two tables you will see that most manufacturers have made some aircraft quite suitable for the mountains and other, very similar models not suitable at all. Usually, but not always, the aircraft capable of getting up to a higher ceiling cost more. Also, you could expect any of the four-seat aircraft to get a couple of thousand feet higher when flown with no one in the back seat and no luggage. Still, you can get into plenty of trouble in these aircraft as well.

The Cessna P210 is the type of aircraft I fly across the continental divide each week year-round. While it is only slightly larger and more powerful than most single-engine aircraft, it is also turbocharged and pressurized and can fly above most weather. Outside the mountains, this general class of aircraft has a higher accident rate than smaller, more basic aircraft. In the mountains, its better climb rate and higher ceiling make it statistically much safer.

Gerrit Paulson photo courtesy SW Aviator magazine.

I will readily admit a love for all the older, smaller-engined two-seat aircraft produced right after World War II as well as the numerous essentially similar modern home-builts. These offer a much less expensive, often much more fun kind of flying than bigger, more expensive aircraft. But remember, much of the fun comes from how low and slowly they go, rather than how high and fast. You *can* fly these aircraft safely in the mountains. Most will eventually claw up to higher altitudes than their service ceilings. Mostly, though, you fly them only on very good days and only down in the valleys.

I will go so far as to recommend that for many pilots, the simplest and least expensive aircraft is the best choice. They are a lot more fun to fly, and because they are inexpensive to operate, you are inclined to take them up on the spur of the moment just because it is a nice day rather than agonize over the cost. When you need bigger load-carrying capacity, or all-weather ability, rent.

Effects of altitude on aircraft engines

Engines produce power by burning fuel and air. For efficient combustion, the air-to-fuel ratio must be very close to 15:1 by weight. This is a lot of air. Reduce the amount of fuel and obviously you are burning less and therefore converting less to heat and power. Less obviously, add more fuel than that to the mix and instead of more power, you get inefficient burning and less power. With less atmospheric pressure at altitude, there is less air available for combustion. You have to reduce the amount of fuel fed into the engine to achieve an efficient fuel burn. The result of less fuel and air being used is less power.

How much less? A lot less! Although there are variations due to compression, camshaft lift and duration, and the like, a fairly accurate rule of thumb says that nonturbocharged *aircraft piston engines will lose at least 3 percent horsepower per 1000-ft elevation gain.*

Assuming 100 percent power at sea level, a typical naturally aspirated (nonturbocharged) aircraft piston engine produces 93 percent power at 2500 ft—close enough to full power to make little difference in flying. At 5000 ft, about the altitude of Denver, Colorado, you might be seeing 85 percent sea-level or rated horsepower. Virtually any production airplane has enough reserve power to fly just fine at that altitude, even if not as well as it does at sea level.

Above 5000-ft elevation, power drops off more substantially. At 8000 ft, there is 76 percent of rated power; at 10,000 ft, 68 percent power; at 12,500 ft, only about 63 percent power. The average small-production airplane flies very poorly indeed at 60 percent of its rated horsepower. At 15,000 ft, less than 1000 ft above the mountain summits, it develops only 55 percent of its rated horsepower. Think of it this way: Your 150-hp engine is now developing only 85 hp. This is like installing the modified snowmobile engine from a single-seat ultralight into a four-place metal aircraft. Would anyone consider doing an engine conversion like that?

This should tell you that on certain days, most small aircraft aren't going to do all the things you might ask of them.

But wait, it get worse!

Density altitude

Altitude—in terms of how high the rocks are—is a constant, expressed in terms of height above *mean sea level*. In terms of how well the engine works and the airplane flies, altitude is really a factor of how thin the air is. High- and low-pressure systems can change the actual pressure of the air at any altitude. Humid air is less dense than dry air. Most significantly of all, *hot air can be much thinner than cold air. Density altitude* refers to what your altitude would actually be if you were experiencing a barometric pressure of 29.92 and a standard air temperature for your altitude.

Let us take a hypothetical walk from my office out onto the ramp. (And let us bring the density altitude chart with us.)

Telluride's altitude is 9087 ft msl, which means its altitude above mean sea level. (Mean sea level is the average of high and low tides.) On a very cold winter day, the *density altitude* may be as low as 7900 ft; that is, the air is as thick as would be found at that elevation on a normal day. Your engine is getting 77 percent of rated horsepower, and almost any small airplane will fly just fine, even if nowhere near as well as it does at low altitude. On a very hot summer day, it is not at all unusual to see the density altitude over 12,000 ft. Now your engine gets only 64 percent of its power. It is just possible that some of the same small aircraft will not be able to get airborne *at all* with a full load. This might be a hidden blessing, because the density altitude required to cross the nearby 14,000-ft mountains might now be something like 18,000 where your engine would be getting 46 percent of its normal power.

1. Determine PRESSURE ALTITUDE as shown below.
2. Find existing OUTSIDE AIR TEMP at bottom of graph.
3. Move upward until you intersect diagonal line indicating
   the current PRESSURE ALTITUDE.
4. Move horizontally to left of graph and read DENSITY ALTITUDE.

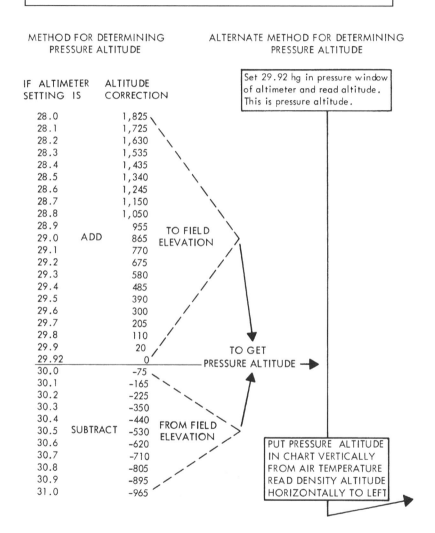

METHOD FOR DETERMINING
PRESSURE ALTITUDE

ALTERNATE METHOD FOR DETERMINING
PRESSURE ALTITUDE

Set 29.92 hg in pressure window
of altimeter and read altitude.
This is pressure altitude.

IF ALTIMETER SETTING IS	ALTITUDE CORRECTION
28.0	1,825
28.1	1,725
28.2	1,630
28.3	1,535
28.4	1,435
28.5	1,340
28.6	1,245
28.7	1,150
28.8	1,050
28.9	955
29.0	865
29.1	770
29.2	675
29.3	580
29.4	485
29.5	390
29.6	300
29.7	205
29.8	110
29.9	20
29.92	0
30.0	-75
30.1	-165
30.2	-225
30.3	-350
30.4	-440
30.5	-530
30.6	-620
30.7	-710
30.8	-805
30.9	-895
31.0	-965

ADD

SUBTRACT

TO FIELD ELEVATION

FROM FIELD ELEVATION

TO GET PRESSURE ALTITUDE →

PUT PRESSURE ALTITUDE
IN CHART VERTICALLY
FROM AIR TEMPERATURE
READ DENSITY ALTITUDE
HORIZONTALLY TO LEFT

Similar charts to the Koch chart reproduced here are provided by the
Federal Aviation Administration (FAA) and can be found on the wall
in the FBO of many airports. ATIS or AWOS information you receive
over the aircraft radio prior to departure now almost universally includes the density altitude.

DENSITY ALTITUDE CALCULATION CHART
Courtesy of the Colorado Division of Aeronautics

Airport Elev. _____ Altimeter Setting _____

Pres. Alt. _____ Temp. _____

DENSITY ALTITUDE _____

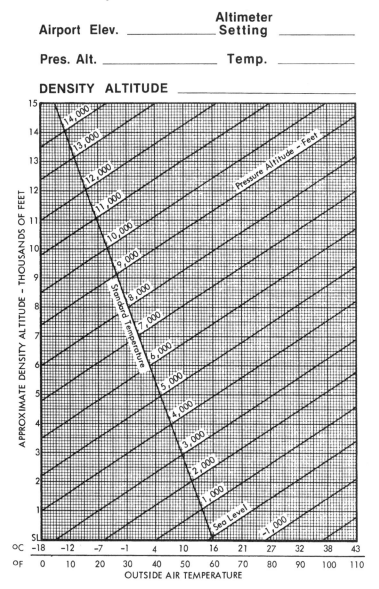

To calculate takeoff distances, use the density altitude, not the airport elevation, and refer to the aircraft operator's manual.

Consulting your aircraft's pilot operator's handbook will now allow you to calculate takeoff distances for various altitudes. But you have to make the calculation using density, not absolute, altitude.

Fuel mixture management

Next to the throttle knob or throttle lever is the mixture knob or lever. By convention, this is or should be red. Pulling it back reduces the amount of fuel in the fuel-to-air mixture going into the engine. Below 3000-ft elevation, about the only time you use this knob is when you pull it out to shut off the engine. At least in a small, low-power airplane, you run full rich at those altitudes almost all the time.

At higher altitude, that won't work. At the very best, the resulting fuel-rich mixture wastes fuel while producing less horsepower. Run a little longer, it will foul the spark plugs, and the engine will still lose more power until it quits. The engine will run rough. When you suddenly give it full throttle, the spurt of extra fuel could be enough to flood the engine and kill it.

Everyone has watched this many times in the mountains. The pilot on final approach to land pushes the mixture knob to full rich—exactly what he or she was taught to do, exactly what the landing checklist says to do, and unless the pilot read the fine print, just what the operator's manual says to do as well. The pilot lands, rolls out, and then adds a little power to turn off at the next taxiway…and the engine just coughs and dies. The pilot can't get it to restart after several tries and has to get a tug to tow the craft clear of the runway. This might be very funny if all she or he had to do was taxi, but less funny if the pilot needed to go around. The fuel-to-air ratio at full rich at a high-altitude airport is far too rich to produce full power, smooth running, or, on some days, even run the engine at all.

Next the same hapless pilot gets ready to depart. But wait. The engine can't even be started! It just produces lots of black smoke but never catches. This is just as well, because even worse occurs if the engine does start but doesn't run very well. The pilot incorrectly assumes it will clear up and takes off anyway. Now the engine is running very roughly, spewing black smoke and producing very little power. With luck, the already extremely long high-altitude takeoff roll is even longer, but somehow the airplane barely climbs out. Everyone aboard was staring death in the face. With no luck, the pilot fails to abort the takeoff, runs off the end of the runway without getting airborne, and hits something. The too-rich-running engine might produce only two-thirds the power it is capable of. This in turn is only two-thirds of the only 60 percent power available at higher altitudes. That is, it is running at 40 percent power. Whoopie! Even in the Denver area at altitudes of only 5000 ft, accidents of this nature are not uncommon, especially in summer.

At Leadville, Colorado, on a hot summer day, pilots with already heavily loaded small aircraft sometimes fill their fuel tanks all the way up, take off, and drop out of sight off the end of the runway. It is a good thing they are headed down the valley toward lower terrain!

In 1999 a pilot of a Piper Arrow passed less than 100 ft over the roof of my house, spewing black smoke. My house is 1^1/$_2$ mi east of the Telluride airport, deep in the valley floor, 300 ft lower than the runway. Somehow he managed to get turned around at treetop height in the narrow box canyon and then began swearing at the line crew on the radio. He accused them of filling his fuel tanks to the top rather than to the middle of the filler necks an inch lower. No, ace, that wasn't the problem. That would only add about 20 lb more to your total weight.

How can one avoid this trouble?

Read the operator's handbook! Generally speaking, when starting a nonturbocharged engine at high altitude, you follow the manufacturer's recommendations, except that you pull the mixture control about one-third of the way out for an estimate of proper mixture. Trial and error will teach you the specific position that works best for your aircraft at any given altitude.

During engine run-up before takeoff, you must set the mixture. Each engine will have a specific procedure that should be in the manual. For most small aircraft, you run the engine up to 1700 rpm, lean the mixture until the power drops, and richen it back up with something like five clicks of the mixture knob back forward until the engine runs smoothly. If you are not sure this is the exact procedure for your aircraft (and if you are reading this for the first time, then you *aren't* sure), you can do a full-power run-up. Set the brakes, go to full throttle, and slowly adjust the mixture to the setting that produces the highest rpms.

With a constant-speed propeller, the 1700-rpm method still works fine. At full power, though, you adjust for highest manifold pressure; or, better, adjust for an exhaust temperature 50°F rich of peak, just as you would in flight.

After takeoff, once you are safely high enough to start screwing around with things, you once again lean the mixture until the power drops, then richen it up until the engine runs smoothly again, plus just a smidgen more. If you have an exhaust gas temperature gage, for most engines, the procedure is to lean out the mixture until the exhaust temperature peaks, then richen it to produce a temperature 50°F cooler. With a fuel flowmeter, the manual will provide you with an optimum fuel flow rate per horsepower, rpms, and altitude and a method of setting optimum mixture.

In all cases, the aircraft's pilot operator's handbook is the authority on specific engine mixture procedures.

Since air pressure decreases with altitude, you will have to reset the mixture for each 1000 or 2000 ft of climb and reset it for a final time at cruising altitude.

On the descent, you will have to enrichen the mixture again. Do not go to full rich on final approach! Just reset for the airport or pattern altitude. Consider this case: You cross the Front Range of the mountains at above 15,000 ft, with the mixture set at about 60 percent of full rich. You then descend into Denver and the airport at about 5000 ft, where you will need about 80 percent mixture. You will have to reset the mixture on the way down, but you will not have to richen it very much.

Turbocharging

Wouldn't it be wonderful, wouldn't it be safer, more convenient, just plain easier if your engine were developing full power? Yes, indeed it would, and systems which do just that have been in regular use since the 1930s. These systems come under the broad classification of *forced induction*. You can achieve sea-level horsepower, or even more than sea-level horsepower, by forcing more air into the engine. Then with more air available, you can burn more fuel and produce more power. A gear- or belt-driven air pump doing this is called a *supercharger*. Much more common for aircraft use is an assembly of two centrifugal or *turbine* fans encased back to back on a common shaft. The exhaust coming out of the engine spins one fan. That in turn spins the other fan on the common shaft which forces air into the intake side of the engine. The engine is then breathing sea-level pressure or even higher-pressure air and developing proportionally greater power. This device looks vaguely like giant Siamese twin snails and is called a *turbocharger*.

Still more power can be produced by passing the compressed, but therefore heated, air through a cooler (like a small air radiator) which cools it so that more expansion will occur when combustion takes place inside the engine. This device is called an *intercooler.*

This extra power is virtually free. Why, then, are not all aircraft engines built this way? Well, it is free, but not exactly *free.* It can cost a significant amount of money to use this free energy. To be sure, the turbocharger itself is precision made to very high tolerances and is therefore expensive to build. More importantly, it is most efficient if designed to spin at unbelievable speeds. It is run by exhaust gases at temperatures well above 1400°F, so it is very hot—turbochargers actually glow cherry red at night. Thus it wears out and requires expensive maintenance. You cannot force more power out of an engine without putting greater strain on it. That in turn means the engine's moving parts wear out sooner. The extra load and power produce greater heat in the engine. Parts expand more with greater heat and then contract without it. Things break or crack more often. More maintenance is required. Much more careful engine management is required of the pilot. Finally, you produce more power by burning more fuel. There is no other way to do it.

For piston aircraft intended to operate above 20,000 ft, a turbocharger is a virtual necessity. Turbocharging is a marvelous and very desirable feature in any mountain aircraft, but for most pilots, the operating and maintenance costs outweigh the benefits. Most small single-engine charter operators could really benefit from having a turbocharger, but they don't. The bottom line is the bottom line, and they can't justify the expense.

For virtually all basic four-seat or smaller aircraft, turbocharging is simply not available, and you are going to have to live without it.

Engine cooling

Most people make the assumption that because air is cooler at high altitude, aircraft engines will run cooler as well. This is unfortunately not the case. Aircraft engines are cooled by heat radiating off the engine being carried away by cooling air blowing over the cylinder cooling fins. At high altitude there is less cooling air blowing over the engine. With a naturally aspirated engine, this is not necessarily a big problem because the engine is developing less power and therefore producing less heat. With a turbocharger force feeding air into your engine, you are getting full power and producing full heat, but you have less cooling air to take care of it.

Moist air exchanges heat better than dry air. High-altitude air is normally drier than lower-altitude air, so engine cooling is again worse than you might hope.

Finally, the nature of mountainous terrain demands prolonged climbs at full power. Departing west out of Denver at a low 5000 ft msl, you are faced with a continuous maximum rate climb to 15,000 ft msl to clear the Front Range. In most small, low-powered aircraft, that would be a climb of at least 30 min, if not a great deal longer as you struggled up the last 1000 ft. Perhaps your engine would not really begin to overheat during a 5-min climb; but on a hot, dry summer day, it is likely you would have to climb at a flatter than normal attitude to go faster and cool the engine better. Add the typical downdrafts always found whenever you need to climb, and now the 30-min climb might be stretching on toward 45 min. In fact, any Denver-area-based small-aircraft pilot can recall at least one day when it was simply impossible to climb over the Front Range of the mountains at all.

Effects of altitude on aerodynamics

Wings produce lift because the air flowing over the top of the wing travels a longer distance at a higher speed. According to Bernoulli's principle, faster-moving air is lowered in pressure. The difference in pressure between the air above the wing and the air below the wing sucks the wing upward. When the air is thinner, the wing has to move faster through the air to produce the same pressure differential needed to produce the same amount of lift. If you are flying at an altitude of 12,000 ft, your speed through the air must be about 20 percent higher than that at sea level to produce the same amount of lift.

This fact is generally realized, but not always understood. The reason is that the speeds you fly are normally read off the airspeed indicator. The airspeed indicator in turn works by reading the pressure rammed into a forward-facing pitot tube sticking out into the airstream. At higher altitudes, thinner air is being rammed into the airspeed indicator. The airspeed reading on the airspeed indicator is in error. If you are flying at an altitude of 12,000 ft, your true airspeed is roughly 20 percent higher than your indicated airspeed.

Wait! Didn't we just see that same figure one paragraph back? The airspeed indicator error exactly matches the extra speed needed to fly at altitude. Thus, even though you are in fact traveling faster over the ground, *at any altitude, the indicated airspeeds you should fly for all aspects of flight are the same.*

The mistake people usually make in this regard is to believe they have to add extra speed for a high-altitude landing. You do have to fly a faster *true airspeed* approach and landing, but the *indicated airspeed* is the same as that at sea level. If you normally cross the fence coming in to land at 65 knots at sea level, then cross the fence at 65 *indicated* knots when coming in to land at 8000 ft. Don't cross the fence at an indicated 77 knots. Your true speed over the ground is already 77 knots when your indicated speed is 65 knots. Would you find it distracting to do these calculations while trying to land the plane? Then don't! Just fly the same indicated airspeeds found in the operator's handbook that you always use.

The other time that this presents a problem is in a climb. The optimum climb speeds are close to, but not the same as, those at sea level. (We discuss more about this later, and you will read it in the operator's handbook for most aircraft.) Because the engine produces much less power at high altitudes, climbs are nowhere near as steep and takeoff rolls are much longer. Sea-level pilots tend to expect the aircraft to deliver performance it is not capable of. They pull the nose off the ground too early and try to climb too steeply. The result is an aircraft flying very slowly, very close to a stall. Again, *fly the indicated airspeeds!* And be very patient while waiting for them to build up.

Effects of altitude on aircraft performance

You can conclude from the last couple of pages that the effect of altitude on aircraft performance is generally negative. This is largely, but not entirely, true. It is absolutely imperative that you study your aircraft's performance charts in the operator's handbook in order to see how poorly your aircraft will perform at the altitudes needed for flight in the mountains.

Before you look at this in detail, consider just a couple of examples.

The Cessna 172 is the most produced aircraft of any type in history, and it is by far the most common type of basic four-seat aircraft in the world. It enjoys the most admirable safety record of any general aviation aircraft ever produced. I own one and use it as a basic trainer. I have towed gliders with one. I use one for charter flights. I swear by this aircraft and would recommend it to anyone.

Nevertheless, at an altitude of 10,000 ft at standard temperature and pressure, a fully loaded early Cessna 172 with a Continental engine has a climb rate of only 230 ft/min. But, you say, that is the old 145-hp engine. The newer ones have a 150-hp Lycoming. Then you gain about 5 ft/min of climb. With the M model, the power goes up to 160 hp, but the fuselage grows a little, the landing gear changes, and the climb *goes back* to only 230 ft/min.

After the Cessna 172, the next most popular basic four-seater is the Piper PA 28 Cherokee/Warrior. At 10,000 ft on a standard temperature day, depending on the specific version, a 160-hp PA 28 Cherokee with a full load will climb at well *under* 160 ft/min! At least one variant will climb at less than 150 ft/min. There are 140-hp variants which climb at only about 100 ft/min at this altitude.

The most popular two-seat small aircraft in the United States is the ubiquitous Cessna 150. At most low-altitude flight schools, because of its very low operating costs, it is the basic trainer of choice. But this is not so in the mountains. At the same 10,000 ft on the same 32°F standard temperature day, a Cessna 150 with a full load climbs at a meager 285 ft/min.

This 10,000-ft altitude was not selected at random. The traffic pattern altitude at Telluride or Leadville, Colorado, is higher than that. On a hot summer day, the density altitude at Aspen, Telluride, Steamboat Springs, or Leadville, Colorado, is higher than that *on the ground.*

Remember that standard temperature at 10,000 ft msl is 23°F. You could vaguely guess that for aircraft in the performance ranges we are discussing here, you could subtract 20 ft/min climb for each 10°F temperature above standard temperature. Now on a midsummer day at pattern altitude in Telluride or field elevation at Leadville, the fully loaded Cessna 172 climbs at 130 ft/min, the Cessna 150 at 185 ft/min, and the Cherokee at about 50 ft/min—that is, not quite enough climb to clear a high fence at the end of the runway. Ask yourself, Is it safe to take off over mountainous terrain with that kind of performance? Get out the operator's manual. On a typical summer day, *none* of these aircraft can reasonably be expected to climb with a full load to the elevations of many major mountain passes, much less the summits of the mountains.

The Koch chart, included as an illustration, gives you a rough calculation for high-altitude performance loss.

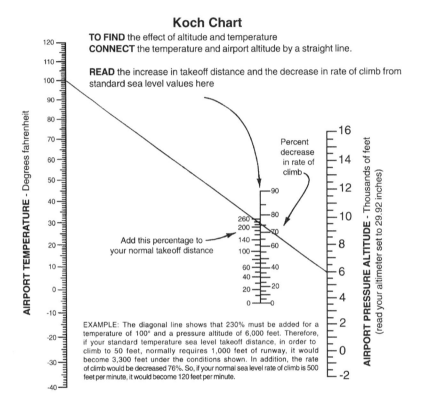

Koch Chart

TO FIND the effect of altitude and temperature
CONNECT the temperature and airport altitude by a straight line.

READ the increase in takeoff distance and the decrease in rate of climb from standard sea level values here

Percent decrease in rate of climb

Add this percentage to your normal takeoff distance

EXAMPLE: The diagonal line shows that 230% must be added for a temperature of 100° and a pressure altitude of 6,000 feet. Therefore, if your standard temperature sea level takeoff distance, in order to climb to 50 feet, normally requires 1,000 feet of runway, it would become 3,300 feet under the conditions shown. In addition, the rate of climb would be decreased 76%. So, if your normal sea level rate of climb is 500 feet per minute, it would become 120 feet per minute.

AIRPORT TEMPERATURE - Degrees fahrenheit

AIRPORT PRESSURE ALTITUDE - Thousands of feet (read your altimeter set to 29.92 inches)

This chart indicates typical representative values for "personal" airplanes. For exact values consult your airplane's flight manual. The chart may be conservative for airplanes with supercharged engines. **Also remember** that long grass, sand, mud, or deep snow can easily double your takeoff distance. Koch Chart courtesy of the Colorado Division of Aeronautics.

Most mountain flight schools use four-seat Cessna 172's as basic trainers, but fly them as *two-seat* aircraft. When it is flown *as a two-seater,* a climb rate of 380 ft/min on the same 10,000-ft 23°F day can be expected. This is fine for airports as high as Aspen, with an 8000-ft field elevation. Indeed, the lowest-performance trainer arguably produces the best pilot in these circumstances.

At Telluride with a field elevation over 9100 ft and a sometimes difficult crosswind situation, we decided after a couple of years to upgrade our 172 trainer with an Air Plains 180-hp engine conversion. Ursula Gilgulin, at Leadville, Colorado, at over 9700 ft, teaches in a Cessna 172XP upgraded to 210 hp. Either of these more powerful versions dramatically outperforms standard production aircraft of the same variety but still falls far, far short of the performance of a standard 172 at sea level.

So if you have one of these bigger engines, you're fine, right? I overheard a visiting pilot ask Ursula this question one hot summer day on the ramp at Leadville. Ursula's reply was that if you had a climb rate of 3000 ft/min, you should have no problem taking off and heading directly east toward Denver. If you had a climb of anything less, maybe you would be better off still heading down the valley just as everyone else does.

At sea level, climb rates of 645 ft/min at gross weight and 1085 ft/min flown solo can be expected from a Cessna 172—and many hotshot sea-level pilots will tell you this is an underpowered low-performance dog of an aircraft! At 15,000 ft *all* small aircraft are underpowered and low-performance dogs.

Again, 10,000 ft is around pattern altitude for many mountain airports. The 15,000-ft msl altitudes are needed to clear nearby mountains. The pilot who just soloed and got a license in Florida now loads a couple of friends into the same aircraft and heads for a vacation in Colorado. At 15,000 ft on a standard temperature 5°F day, the very same plane he soloed in with a better than 1000 ft/min climb now only sees 22 ft/min. In summer, the climb rate may have reached zero well below 13,000 ft. *Flown as a two-seat aircraft,* the same Cessna 172 only manages a 150 ft/min maximum climb at 15,000 ft. Any day, summer or winter, flown as a four-seater, the Piper Cherokee PA 28-160 will never get anywhere near 15,000 ft.

Needless to say, this is not enough to outclimb the 400 to 500 ft/min downdrafts which can be encountered anywhere in the mountains almost any day of the year, much less the more than 1000 ft/min downdrafts all mountain pilots have encountered.

This is what gets sea-level pilots in trouble during their first flights in the mountains. They trained and got a license in an aircraft with a climb rate of better than 1000 ft/min. Now, *in the very same aircraft,* the climb rate is a useless 22 ft/min, and very high jagged granite peaks are in their way.

This is also most of the explanation of why mountain-trained pilots get in less trouble. Every time they have gone flying, they never experienced anything better than these marginal climb rates. They just don't know any better, so they have always had to deal with this problem right from their first introductory flight. That is all they expect the aircraft to do, and they fly it accordingly.

Speeds to fly change with altitude

Climb speeds. In most pilot operator's handbooks, there is a speed for V_x, best angle of climb, and V_y, best rate of climb. At increasingly higher altitudes, these two speeds grow closer together until they meet. At that point, it may well be that the resulting speed is the *only* speed you can fly without losing altitude, as the rate of climb has now reached or will soon reach zero. This is the aircraft's absolute ceiling.

Choosing again the manual for a Cessna 172, but expecting generally similar results from all small aircraft, we find the following: At sea level at full gross weight, the best rate of climb is at 80 indicated mi/h. The best climb angle is at 70 indicated mi/h. (Earlier Cessna airspeed indicators read in miles, not knots—no doubt a decision made by the marketing department because the bigger numbers sound faster.) At 15,000 ft, the best rate of climb has dropped to 76 indicated mi/h. The best climb angle has increased to 72 mi/h. Since the rate of climb is now only 22 ft/min, it is apparent that you will not be flying any speed outside the 4 mi/h spread between those figures for very long. Compared to your sea-level cruising speed in that aircraft of 130 mi/h, it is equally apparent that you are in for a fairly prolonged trip.

Given the relatively low rates of climb at high altitude, you might correctly conclude that nearly all your flight might consist of a prolonged climb at the best climb rate with much circling to clear high terrain, followed by cruising flight at about the same speed, finally ending with a very brief descent to your destination airport. This is still much faster than driving and much more enjoyable.

Optimum cruise speed and altitude. Most nonturbocharged small aircraft have a relatively low optimum cruising altitude. This altitude will vary somewhat with engine type and aircraft design, but typically will be encountered about where full throttle produces the maximum allowable percentage of horsepower for cruise at the maximum allowable cruise engine rpms. Does this sentence make any sense to anyone? Consider a couple of examples.

Consider the ubiquitous Cessna 172. Maximum continuous cruise power is limited to 75 percent of rated horsepower. Maximum sustained rpms are limited to 2500 at sea level, a little more at altitude. You achieve your fastest *cruising* speed at about 7500 to 8000 ft. Full throttle at that altitude produces only 75 percent of rated horsepower. The engine is turning an acceptable 2600 rpms. Cruising speed is a true 130 mi/h (or an indicated 118). Fuel burn is around 7.7 gal/h, resulting in a range of 600 mi and an endurance of 4.4 h.

At sea level, you have a faster *maximum* speed of 138, but you can't *cruise* at that power setting for more than a couple of minutes. You have to throttle back to the maximum allowable 2500 rpms. Compared to flying at 9000 ft, your speed would then be 5 mi/h slower at 125, while your fuel burn would be nearly 1 gal/h higher, giving you a reduced endurance of only 4 h and a reduced range of 530 mi.

At 12,500 ft—a reasonable cruising altitude if you flew up the major valleys between the mountain passes—your economy and endurance are improved at the cost of speed. You could cruise at full throttle at only 7.2 gal/h, but your speed is back down to 125 mi/h. Your endurance increases to 5 h and your range to 630 mi.

For the compulsively economically minded, reducing the rpms to only 2200 at 12,500 ft slows the aircraft down to a cruising speed of only 100 mi/h (which is still a lot faster than your car, or at least faster than my car) while reducing fuel burn to 5.4 gal/h and increasing the range to 670 mi, that is, 140 mi extra of range compared to normal sea-level cruise.

Contrary to what intuition might imply, this kind of babying of the engine does not in fact promote longer engine life. Normal cruising does. Exactly why, then, would you choose this last option?

We once purchased a glider from Driggs, Idaho, and flew up from Telluride, Colorado, to get it with our 180-hp Cessna 172, a trip of about 3.5-h duration, which left us with a reasonable 12-gal fuel reserve. On the trip back we were limited by the glider's maximum tow speed of around 85 mi/h. Add the benefits of a strong tailwind. Subtract for having to maneuver considerably around weather. Fudge in something for some higher than planned elevations to keep the glider within safe gliding distance of acceptable landing fields, and the whole trip took a continuous 5 h. Despite towing a heavy two-seat glider, we arrived home with a reserve of 6 gal. Flying faster would have required a fuel stop and produced a longer trip.

More airplane is better

In looking over the effects of altitude on small aircraft performance, one is struck by how truly marginal the performance of most small aircraft becomes at altitudes much lower than that of most major mountain passes, much less summits, of the Rocky Mountains. The resulting statistics are just what you would expect. *Four-seat aircraft of less than 180 hp are very significantly overrepresented in mountain flying accidents.* Baker and Lamb cite inability to outclimb terrain as a factor in 49 percent of accidents for this type of aircraft versus only 14 percent for more powerful aircraft. Less intuitively perhaps, but at least equally significantly, this is a cause of the accident 6 percent of the time when these aircraft are flown solo, 13 percent of the time with two occupants, 24 percent of the time with three, and 30 percent when flown with four. Go back to the rate of climb tables to see why. A fully loaded PA 28-160 cannot reach within 2000 ft of a density altitude of 15,000 ft at all. A fully loaded C-172 climbs at only 22 ft/min. The density altitude at major mountain passes on a July afternoon is normally at least that high. *A prevailing wind of less than 1 mi/h produces a downdraft in excess of 50 ft/min on the lee side of the pass!* In fact, it probably would produce a downdraft of closer to 100 ft/min.

Fly the same four-seat Cessna as a two-seater, and your climb rate goes up from 22 to 155 ft/min, which is not great but better enough to reduce the accident rate for that type of accident from 30 to 15 percent!

The ability to climb is a major factor is mountain flying safety. A more powerful aircraft with a higher rate of climb and higher ceiling offers a greater safety margin. A couple of personal anecdotes will illustrate this.

"There I was!" All good personal anecdotes begin this way.

So *there I was,* coming back from Denver to Aspen, Colorado, flying a Cessna 172. The total load was me, my pregnant wife, and fuel tanks about two-thirds full. Westerly winds of about 20 knots were predicted at 18,000 ft. We went by the summit of Mount Evans east of the continental divide at over 15,500 ft msl, looking safely down at the 14,000 ft msl we wanted to clear the 12,000-ft msl Loveland Pass. Then in less time than it takes to read this, everything loose in the plane was bouncing around the cabin, the rate of climb meter was pegged at 2000 per minute down, and we had already turned around down Clear Creek Canyon toward Jeffco Airport. It just didn't seem that important to get home just at that particular time. We flew home late the same evening without incident. If I return to this theme again later in the book, it is because it is *not* an unusual kind of occurrence.

Everyone who flies small aircraft in the mountains has had multiple experiences like this one.

But *there I was* again, 3 or 4 years later, flying as corporate pilot with Telluride businessman and personal friend Grant McCargo, waiting at the Jeffco Airport and listening to an endless stream of pilot reports of moderate to severe turbulence, inability to climb over the mountains in downslope Chinook winds, and requests to return to the airport. But there was a pattern developing. All the turbulence was being reported west of interstate 25. We took off in surface winds gusting over 30 knots with winds aloft blowing the wrong way at over 70 knots. But we climbed northeast, *away* from the mountains, and circled way out over the plains east of Denver International Airport's B airspace. There was certainly some turbulence, but turbulence in a 4000-lb gross weight aircraft like the Cessna P210 is far more comfortable than the same turbulence encountered in a 2300-lb gross weight Cessna 172.

We then headed into the mountains at an altitude of over 19,000 ft msl. With our load that day we might have expected the turbocharged, fuel-injected, intercooled Cessna 210 to maintain a climb of better than 500 ft/min. Instead, we were losing nearly that amount. But no problem. We still crossed the Front Range at over 17,500 ft, well higher than most of the Chinook downslope winds. With a headwind of 75 knots over the mountains, we still maintained a ground speed of 99 and were home in time for dinner. At our altitudes, we encountered only very light turbulence on the climb and no turbulence at all en route, as opposed to the severe turbulence people were reporting below us. In the same Cessna P210, I have flown back home at altitudes of 24,000 ft to get above weather and pushed through head winds of over 85 knots in uncomfortable but certainly survivable turbulence.

A Cessna P210. Outside the mountains, the increased complexity of this general class of aircraft, such as its retractable landing gear, variable-pitch propeller, pressurized cabin, engine-cooling flaps, fuel management, and more, translate into more accidents than with more basic aircraft. In the mountains, this aircraft's better climb ability and higher ceiling give it a statistical safety advantage over simpler aircraft. Gerrit Paulson photo courtesy SW Aviator Magazine.

The 310-hp turbocharged, pressurized 4000-lb airplane can't make the trip every single day, but it can make it more often than the 2300-lb, 180-hp plane and do it with far less drama. A 19-seat, twin-turbine Beech 1900 airliner can usually make the trip when the single-engine Cessna 210 can't; and on a few of the days the Beech 1900 isn't flying, 200-seat Boeing airliners are cruising overhead with flight attendants walking down the aisle with the beverage cart. Of course, a Boeing is far too large to land at virtually any mountain airports.

Given the same airplanes on different days, there were no strong winds and no turbulence, but...

So there I was. Now instead of a pregnant wife, I had a wife and a 1-year-old child headed home from Denver to Telluride. Taking an indirect route to avoid high terrain, we were soon headed into unpredicted towering black masses of freezing rain over South Park. You just don't fly into known ice in the mountains in a small airplane with or without a baby on board. Even with supplemental oxygen, which a 1-year-old child would refuse to wear anyway, 17,000 ft would have been our maximum altitude—nowhere near enough to climb over the weather. Despite being only 7 mi out from the Buena Vista, Colorado, airport, we were forced to backtrack 60 mi, then fly west over 100 mi to Eagle, Colorado. There we refueled and waited out a few cells before heading south over McClure Pass. Sure enough, although we were only 5 mi out from North Fork Valley airport in Paonia, Colorado, once again we were blocked by freezing rain and backtracked 60 mi to land in Glenwood Springs and wait out more storms before choosing yet another route. We landed in Telluride at 7:30 p.m., a total trip time of 9 h, or only 1 h *longer* than it would have taken to drive in a car!

My personal record flight in a small aircraft from Denver to Glenwood Springs, Colorado, west of the Continental Divide is longer than 4 days because of weather.

But as we were coming home in the pressurized Cessna 210, our departure was delayed more than an hour by a late passenger in bad weather predicted to go worse. Already in the clouds less than halfway home at 18,000 ft, we began taking very light ice. We requested and were given first 20,000 and then 22,000 ft to get above the icing. Had we needed to do so, we could have climbed up to 24,000 ft. On about one-half of our winter trips to Denver in that aircraft there are clouds and snowstorms below us in the mountains, but we are flying above the weather in the clear, bright sunshine. On about one-quarter of our winter flights from Telluride to Denver we have to make an instrument departure or instrument approach, or both, through the clouds. We still couldn't see anything out the windshield that day, but the icing stopped and the small accumulations we had quickly sublimated off even in the clouds. We were comfortably warm in the pressurized cabin in our shirtsleeves. The passengers in back could converse easily without headsets and so had no idea where we were or what we were doing. The instrument approach into Telluride got us to within a mile of the end of the runway with nothing in sight beyond the cowling. We then diverted to Montrose, Colorado; broke out of the clouds with 2000 ft to spare, and got a $1^{1}/_{2}$-h airport shuttle ride home. Instead of our normal 1 h 15 min-long trip, including the car ride it took 3.5 h.

We still got home by the dinner hour, 4 h faster than it would have taken to drive, even if the roads had not been closed by the blizzard. This was certainly much faster than it would take to get to Denver International Airport 2 h early for the security check for an airliner that ended up having to divert to the same airport we did.

One more time

In February 2002, the weather in Telluride was fine and the weather in Denver was fine. While there was a major storm system moving into the region out of eastern Oregon, the mountains were predicted to be very acceptable VFR with no airmets. You know that isn't going to be true—why else would I be recounting this incident?

Sure enough, scarcely 15 min out of Telluride, we could see that the weather over the Collegiate Peaks along the Continental Divide was nothing like predicted. In my Cessna 172, I would have turned around without even pausing to consider exploring further. In Dean Gianpietro's turbocharged Bonanza we just called Denver Center for an instrument clearance and began climbing. We climbed into an open layer between clouds at 19,000 ft. The descent into Denver was in clear air. After lunch for the return trip, things looked worse as high winds began producing turbulence all over the region. We accordingly filed an instrument flight plan, climbed through a very few clouds to 20,000 ft, and headed home in bright sunshine and no turbulence above a solid undercast. By the time we neared Telluride, we had to descend through only a very few clouds above 14,000 ft and bump through very light turbulence to land. The unexpected weather which would have prevented any consideration of getting to the destination in a smaller plane was scarcely even a factor beyond some additional radio communication and the inconvenience of wearing an oxygen mask.

In a small four-seat, low-performance airplane, the entire 2-h flight from Telluride to Denver is in, below, and dodging around 14,000-ft mountains, down in and among the clouds. In a six-seat turbocharged airplane, only the takeoff and climb-out truly need be in the mountains except in extreme weather. Thereafter often you can expect to simply fly at a much higher altitude well above the mountains and most of their weather. The trip takes half as long. There are obvious safety and convenience factors involved. Yet even in higher-performance aircraft, even in large airliners, mountain flying skills and knowledge have a role to play in the safe outcome of the flight.

Worldwide, statistically, aircraft like the turbo Bonanza or turbo Centurion have many more accidents than the simple Cessna 172. Possibly this is because they are far more complex and require more systems management than the simpler aircraft. Many Bonanza and Centurion accidents result from something no more complicated than running out of fuel in one tank, even though another still has plenty of fuel. In the mountains, these types of aircraft are safer simply because they can climb so much higher. In the mountains, the overall statistically safer, less complex aircraft such as the Cessna 172 is potentially more dangerous because it cannot climb as well or climb high enough to fly above weather.

In the mountains, safety with either aircraft is largely a result of pilot decision making. I make no excuses for the fact that I have the most fun making the flight in the lightest, most basic airplane in the most favorable possible weather. But the important word is *weather*. The smaller and more basic the aircraft, the less often you can safely fly it in the mountains.

Effects of altitude on pilots

We have examined the effects of reduced air pressure and reduced available oxygen on small aircraft. Let us now turn our attention to the effects on human beings.

Hypoxia

Your body is powered by the food you eat and the water you drink, metabolized with the oxygen you breathe. Although there is plenty of oxygen at the altitudes that a small aircraft can reach, that oxygen is under much lower pressure than at sea level. Your body requires pressure to pass the oxygen through the alveolae in your lungs and get it into your bloodstream, where it bonds with hemoglobin and goes where you can use it. Available oxygen is substantially reduced at higher altitudes. At some altitude, which is slightly different for all individuals, your body will not be able to take up sufficient oxygen to function properly. At very high altitudes, there is not enough available oxygen to sustain life, and you begin to die. How long this takes is again a factor of available oxygen. There is plenty of oxygen at high altitude, but because it is at reduced pressure, your body cannot take enough of it in.

In pressurized jet aircraft, one of the biggest concerns is how very rapidly a pilot can lose consciousness because of reduced oxygen. While the cabin is pressurized, the occupants of the aircraft are living in air pressurized to whatever equivalent altitude the cabin pressurization is adjusted to maintain. Typically, this is some altitude lower than 9000 ft. Should the pressurization system fail, either gradually or catastrophically if a door or window failed, useful time of consciousness would then be reduced to whatever would be the case at the altitude the aircraft was actually flying.

An average person will pass out in 15 to 20 s at an altitude of 40,000 ft. At 30,000 ft where most airliners fly, a person might last a minute or two. Aircraft that fly at these altitudes must be equipped with emergency oxygen systems which have masks that automatically deploy with reduced cabin pressure. These numbers are of only theoretical interest to the pilot of a small piston-engine aircraft.

But look at the things that can happen at altitudes a little closer to home. Pressurized single-engine aircraft such as the Piper Malibus or Cessna P210s will frequently be assigned altitudes of 22,000 ft crossing the Rocky Mountains in marginal weather. Should the cabin pressure fail, the average person would pass out in 10 min. This time frame contains a subtle extra element of danger. If you were on the way to passing out in 1 min, very early into the process you would be acutely aware that something very bad was happening. You might very well guess what it was and descend to lower altitude or reach for the oxygen mask. If the process took as long as 10 min, though, onset might just be gradual enough to go undetected until you actually did pass out.

A local pilot related the story to me of a 1999 flight in a Cessna turbo 182RG from California to Telluride. She was in the right seat, cruising at 19,000 ft eastbound in the clouds. Since she was not acting as pilot, she only occasionally ran her eyes over the instruments while trying to get a little rest. Then in one of these glances she noticed the VOR needle way off center. Looking at the attitude indicator, she observed a shallow banked turn still farther away from the needle and a gradually decreasing altitude. She pointed to the appropriate instruments and nudged the pilot. He was asleep. When there was no response, she prodded him hard in the ribs. There was still no response, so she got the airplane back on heading and then back on route. This awoke the backseat passenger who was also a flight instructor. They discovered that the pilot's oxygen hose to his nasal cannula had caught in his seat belt and pulled out of the fitting. When they reconnected him, he came to in about 1 min and asked what was going on. Presumably, he had been without oxygen for the entire flight and gradually slipped away.

At 18,000 ft—less than 4000 ft above the Rocky Mountains—lower than some Alaskan and Mexican volcanoes, the average person can be expected to pass out in 30 min without supplemental oxygen. Not everyone will, of course.

So, fine, you are thinking to yourself, but my airplane probably can't struggle up to anything over 15,000 ft. I don't expect to stay that high for very long. I don't expect to pass out. What does any of this have to do with me? Plenty.

You might not actually pass out, but you certainly could expect to experience some effects of reduced oxygen in your system. Most of these effects reduce your abilities as a pilot. Various factors other than just altitude alone can cause reduced oxygen saturation in the body. Collectively, these are called *hypoxia,* or lack of oxygen.

We normally think of *hypoxic hypoxia,* which is the lack of oxygen due to the lack of oxygen supply. This is almost synonymous with high altitude. There is also *hypemic hypoxia,* which is the result of an inability of oxygen to bind to hemoglobin due to blood loss, anemia, or toxins in the blood. Pilots of small aircraft take notice! Your cabin heater is a cuff around part of the engine exhaust system. Cold air from outside passes through the chamber, is heated by the exhaust pipe, and flows into the cabin. A tiny crack or a pinhole of corrosion in the exhaust pipe would allow exhaust gas into your heated cabin air supply. Engine exhaust gases contain carbon monoxide, which not only is poisonous, but also binds better with hemoglobin than does oxygen. Carbon monoxide is odorless. You smell nothing, you are not at a very high altitude, but you begin to become hypoxic anyway. Does this sound like something you would really, really love to try, but you just can't manage it because your heater is just too well maintained or it is too warm a day to use cabin heat? Well, you're in luck! The tobacco industry expends billions of dollars a year bribing members of Congress and advertising a product which sells for only $0.10 per dose and brings you enough carcinogens to end your life *plus* all the benefits of breathing engine exhaust! A sea-level-dwelling moderate smoker will begin to feel some of the effects of hypoxia at altitudes as low as 3000 ft. This result produces such demonstrably negative effects on performance that the United States Air Force requires smoking pilots to use oxygen at night at altitudes of only 5000 ft. In considering the effects of smoking for pilots, it must be remembered that if you are already at high altitude, then the effects of second-hand smoke constitute something of an entirely different nature than just the annoyance they do at low altitude, and you should go to some extremes to avoid them, particularly before flying at night.

This can also happen to you if you suffer traumatic injury and go into shock. Blood loss or loss of blood pressure produces hypoxia. I once donated a unit of blood at the hospital at 8000 ft, felt very good about myself, walked out into the lobby, and fainted and fell to the floor.

Histotoxic hypoxia results from tissue poisoning such as may result from various drugs. Again, smokers have the edge on the rest of humanity, because all sorts of things other than just the carbon monoxide found in cigarette smoke cause tissue poisoning leading to hypoxia. Are you, like me, someone who can't take even a whiff of tobacco smoke without feeling nauseous and getting a headache? Still want to try for histotoxic hypoxia? You might try knocking back a cool beer before the flight. Alcohol is another tissue-poisoning substance producing this effect. This is part of the explanation of why people who can drink three beers at sea level become intoxicated after drinking only one beer at a ski resort.

With less oxygen going through your system, things you consume are metabolized more slowly, so their effects can stay with you longer. While the FAA regulations prohibit drinking alcohol for 8 h before flying, 12 h might be a better idea if you will be spending those hours at high elevations.

Marijuana is illegal for flying anyway. You presumably avoid it for other reasons, but it is worth noting that it, too, can be a factor in producing histotoxic hypoxia. Most marijuana smokers will insist that they are unimpaired the morning after smoking. Medically this is patently untrue. Their motor functions are relatively unimpaired, but their judgment is significantly impaired for 2 or 3 days (during which a blood or urine test can readily detect the marijuana, by the way). The combination of the histotoxic effects and high altitude can double this period of impaired judgment. Limited marijuana smoking is generally tolerated at resort communities, and it is possible to occasionally find pilots at resorts who smoke it. In at least two of the examples of dangerous flying cited later in this book, judgment impaired by marijuana smoked the day before was believed by people who were there to have been a contributing factor.

Stagnant hypoxia occurs when decreased blood flow results in a decreased oxygen supply to the tissues. While it is often seen in connection with shock and traumatic injury, pilots could encounter this condition as a result of exposure to extreme heat or cold or high *g* forces.

Susceptibility to hypoxia

Pilots should remember that when flying in the mountains, they are *already* at altitudes where a mild degree of hypoxic hypoxia is always present. Anything contributing to any of the other forms of hypoxia would have the better part of the job already accomplished. A smoker who had a sip of schnapps under the belt on an emotionally stressful, bitter cold day with a leaky heater could well be beyond hypoxia before even taking off.

If you live at sea level but just spent your first night at altitude before making a flight the next day, you have not acclimated much to the altitude. More likely, you have suffered from very mild altitude effects all night, which made sleeping difficult, and are even more susceptible to hypoxia now.

If you have a *cold, the flu, asthma,* or any other form of respiratory ailment, then you have a reduced ability to transport oxygen and are temporarily more susceptible to hypoxia.

Dangers of hypoxia

What dangers does hypoxia hold for pilots? Let us take a look at the symptoms in the order in which they occur. Note, however, in this regard that not all people will experience all the symptoms. The same persons might experience different symptoms and to differing degrees on different occasions, and different people will have very different physiologies and feel the effect at very different altitudes.

Loss of night vision. Very significant loss of night vision occurs with only mild degrees of hypoxia. Remember that the Air Force requires smokers to use oxygen at night at less than 5000 ft. Many smokers are effectively night-blind at that altitude.

Anxiety. This would not be advantageous when you are trying to keep your cool in demanding flying situations. Further, you might be feeling a degree of anxiety anyway without the hypoxia-induced anxiety.

Headache. Now that the statute of limitations has run out, I can admit that in a previous life before learning not to, we sometimes took paragliding passengers up to 18,000 ft in thermals. We thought it was really great. They mostly complained of headaches. (See also *anxiety* and *nausea* above and below.) They seldom tipped on these flights.

Nausea. See above. By the way, an experienced tandem paraglider pilot can usually slide the passenger slightly over to one side to avoid being hit when the passenger gets sick. A nauseous pilot, even with no vomiting, is still clearly impaired.

Dizziness. Together with dizziness, you can expect numbness and tingling around the mouth and nose, and numb and tingling fingers. You can still fly in this state, but it is obviously harder to do it well.

Fatigue. Obviously this is not great when you are trying to fly.

Blurred vision. Note also that fatigue, dizziness, and anxiety can produce this symptom independent of hypoxia.

Slow thinking. What?

Impaired judgment. This is probably the single most dangerous symptom for a pilot in the mountains, since judgment rather than simple flying skill is the most crucial element of safe mountain flight, and poor judgment is less easy to assess than poor skill.

Tunnel vision. Obviously this is serious, and sometimes it is a precursor to blacking out.

Cyanosis.   Your lips and fingernail beds turn blue. The significance of this is that several other serious effects are usually present before blue nails and lips are noticeable.

Mental confusion.   It is difficult to think clearly, coordination is poor, and motor functions are slow. Speech is sometimes slurred. Reactions are slow. The danger of flight in this condition is obvious!

Euphoria.   Wow! It really is fun to fly at high altitude (except that this is just a variation of poor judgment)!

The final stages of hypoxia are *unconsciousness* and *death*.

Concurrent with all the other symptoms of hypoxia is a curious one indeed—the impaired judgment of hypoxia often includes an inability to recognize the symptoms of hypoxia. That is, *one of the symptoms of hypoxia is believing that you are not hypoxic.*

So at least when we are on the ground and not hypoxic, we can all agree that even mild hypoxia produces a degree of impairment that is highly undesirable when flying an airplane. It is impairment rather than discomfort, which is the concern with hypoxia. Since the signs of hypoxia are not always obvious to the pilot, it is only prudent to take steps to prevent it before the flight.

Hyperventilation

Hyperventilation produces some symptoms easily confused with *hypoxia*. These include numbness and tingling of the mouth and fingers and dizziness. In extreme cases, hyperventilation can result in unconsciousness. Hyperventilation is typically stress-related or fear-related. Very rapid breathing causes an excessive rate of exhalation of carbon dioxide from the blood. The result is raised blood alkalinity, which causes the symptoms. The cure for hyperventilation is to force yourself to slow down your breathing. In theory, you could return to normal levels of carbon dioxide by breathing into a paper bag, and that is the normally recommended procedure. The fact is, however, that when you are at high altitude in a small airplane, you don't know whether you have hypoxia or hyperventilation. Since you don't know, I suggest breathing slowly *and* using oxygen. Either way, the symptoms will resolve themselves.

Once they are gone, what should you do then? Want to try an experiment? Remove the mask. If the problem was hyperventilation, the symptoms will be gone. If it was hypoxia, they will come back.

Would you rather put off scientific curiosity for another time and just keep flying? Either way, the symptoms went away, didn't they? Why not just keep wearing the mask? If fear or anxiety was causing the hyperventilation, the very fact that you are doing something about the problem may help alleviate it regardless of the physiological effects of the cure. A little extra oxygen at high altitude is almost always beneficial anyway, even if not always absolutely required.

Avoiding hypoxia

Since hypoxia is the condition of being without adequate oxygen, it follows that both the cure and the prevention involve no more than providing an adequate oxygen supply.

Fly low.   By avoiding the altitudes where hypoxia occurs, hypoxia can be avoided. Above those altitudes, supplemental oxygen is required. Let us review the FAA part 91 regulations concerning supplemental oxygen. (The regulations for part 135 charter and part 121 scheduled air carriers are more stringent.)

- Below 12,500 ft msl, supplemental oxygen is not required.
- Supplemental oxygen must be used by the flight crew if the aircraft is flown over 12,500 ft for more than 30 min.
- Supplemental oxygen must be used by the required flight crew on any flight over 14,000 ft.
- Supplemental oxygen must be available for passengers on flights over 15,000 ft.

Acclimatize

It is very possible to change the physiology of your body to make it perform well at higher altitude. What you want is to increase the number of red blood cells in your body to carry more oxygen and to increase the capillary density in your tissues to deliver those red blood cells more easily. All you have to do to achieve this is to stay at high altitude for a couple of weeks.

It is quite possible to fly the 14,000-ft peaks of Colorado legally without oxygen, and that is precisely what most mountain-based pilots usually do when flying small aircraft. Most of the flight takes place in valleys at altitudes below 12,500 ft. They pop up over that altitude but stay under 14,000 ft when crossing passes and drop back down as soon as possible.

At least that is what they claim to do. Quite likely, the oxygen requirements are the only FAA regulations routinely violated by large numbers of mountain-dwelling pilots. The reason is physiology and acclimatization. A person of average physiology who lives below 3000-ft elevation is effectively acclimatized to sea-level conditions. It is at least statistically likely that such a person would fairly quickly feel the effects of hypoxia should he or she violate the FAA requirements by only a small amount. Mountain resort medical clinics every year see very large numbers of people feeling significant detrimental effects of altitude at elevations well *under* 8000 ft.

On the other hand, people who live at 6000 ft have acclimated to those altitudes. Their bodies have a greater blood volume, more red blood cells, higher capillary density, larger lungs, and more alveolae than would be the case if they lived at sea level. These people cannot go 6000 ft higher than a sea-level dweller without hypoxia, but can certainly go 1000 or 2000 ft higher. In flying terms, they can easily pop up to over 14,000 ft for well over 30 min to cross a mountain pass with almost no ill effects. Someone living in Telluride at over 9000 ft or in Leadville at over 10,000 ft will suffer few ill effects while working all day long at the over 12,500-ft altitudes where the snow cat drivers, ski lift operators, and ski patrol spend their entire workweek doing hard physical exercise. Neither will the helicopter ski guides who are dropped off to work at 13,500 ft need oxygen when sitting calmly on the flight up that they will not require for the athletic ski back down. Someone living at 9000 to 10,000 ft, such as the entire populations of Leadville and Telluride, Colorado, can quite likely—but not certainly—fly around at over 16,000 ft for a few hours and feel relatively few ill effects. The populations of La Paz, Bolivia, Cuzco, Peru, and Kathmandu, Nepal, are born, live out their natural lives, and die well above these sorts of altitudes with little acute difficulty. How?

After a week to 10 days of living at 8000 ft, your body will have produced a much higher volume of red blood cells and you will feel about as comfortable as the natives do. Drinking plenty of water will help this process. After you have lived at this altitude for a year, the density of the capillaries in your body tissues will have increased as well, providing an improved delivery of all the extra oxygen your increased red blood cells can carry. You never make a 100 percent adaptation to altitude. Babies born at the hospitals in Leadville, at 10,000 ft above sea level, or even Aspen or Glenwood Springs, 8000 and 6000 ft above sea level, respectively, are statistically slightly lighter in weight than sea-level babies, even though their mothers are slightly more fit and healthier than the general population.

People who engage in regular vigorous aerobic exercise and are well hydrated will have produced many of the same changes in their bodies that result from acclimatization to high altitude. They will be far less susceptible to the effects of altitude and hypoxia than the rest of us.

The human body is amazingly adaptable. These days, Mount Everest climbers are sometimes hauled up to the summit by teams of Sherpas while breathing about as much oxygen as they do at sea level through masks, with a good supply of more oxygen carried along by more Sherpas. But this was not so with Swedish hippie (and paraglider pilot) Goren Kopf. Goren loaded up his bicycle and a trailer with all his supplies in Sweden and rode down to Nepal. Then he hiked in with everything he needed to base camp all by himself. This would tire many people out, but Goren next scaled 29,029-ft Mount Everest alone with no oxygen, accepting no outside help from anyone. At 29,029 ft, Goren Kopf was at the end of an arduous 7-day ascent with a heavy backpack, following a couple of weeks' hike in on top of a 4-month bicycle ride down from Sweden pulling a trailer. At 29,029-ft elevation, the average person passes out from lack of oxygen in 1 min just sitting still. Goren could not *live* at those altitudes, nor could any other human being, but he could spend a very small number of days there.

Why wouldn't Goren pass out and die as all the rest of us would? The answer is acclimatization. The months he spent getting to the mountain and hauling in his supplies at high altitude produced physical changes in his body. Why would the rest of us pass out? Most of us live below 9000 ft, and the very rapid climb of an airplane to those altitudes leaves out the weeks or months needed to acclimatize to very high elevations. Indeed, most of us would never acclimate to very high altitudes whatever we did. Most of us just don't have the right genetic makeup. Goren could not actually live at those altitudes either—he could only go longer before passing out.

Recovery times from hypoxia vary greatly as well. I recount the story of a Telluride realtor who was flown up to 17,000 ft without oxygen (not by me) to take photographs of some property. At the time, she only felt minor effects, but on the descent she acquired a splitting headache which persisted for several hours. That afternoon she flew an already scheduled flight lesson but found her motor skills had deteriorated to the point where she was essentially incapable of executing a normal safe landing. Yet this was a healthy, physically active, well-acclimated, young person who lived at 9000 ft.

A few years ago after an hour at 19,000 ft on a flight to Denver, I began to wonder if I was feeling just a little light-headed. A check of all the aircraft systems revealed that the cabin pressure switch had been inadvertently bumped into the off position. But I was nowhere close to passing out and not obviously impaired in any way. Grant McCargo, also in the plane, 10 years my junior and a determined high-altitude distance runner, felt nothing whatsoever. After a brief discussion, we decided that since our descent would begin in less than 5 min, it would be better to just leave the cabin pressure alone rather than risk blowing out our eardrums when it came back on. We elected not to use the emergency oxygen system because it is a one-use system rather than a refillable one on that aircraft, and the descent would cure our problem. As far as I know, neither of us had any recovery issues at all.

Why would I do better than the younger, fitter woman in the previous example? All the evidence would predict the opposite outcome. One very major factor might be her childhood exposure to second-hand smoke. Another might just be choosing the right parents—there is a genetic component to this. The truth is, there is really no good way to tell.

Supplemental oxygen

Contrasting Goren Kopf at 29,000 ft with all the rest of us tells us that it is wrong to make any very firm assumptions about just when hypoxia may be a problem. Given that aggravating factors such as cold and stress may well be expected, it is only prudent to act conservatively. *The easiest way to be safe is to stay within the FAA oxygen regulations.*

Here is how you do that:

Pressurized cabin

The simplest form of supplemental oxygen to use is a pressurized cabin. You just set a dial somewhere on the panel for the desired cabin altitude, and you flip on a switch. If your airplane is pressurized, you need read no more on this subject, although you might want to remind yourself about where the emergency masks are located. Most general aviation aircraft expected to operate regularly above 20,000 ft are pressurized. It is the 14,000- to 20,000-ft altitude range where supplemental breathing systems are most commonly used. Unfortunately for you, basic small aircraft do not have pressurized cabins.

Oxygen systems

If your aircraft is not pressurized, either it can be fitted with a built-in oxygen supply or you can carry portable oxygen bottles. Both systems work equally well. The advantage of the built-in system is greater oxygen volume and convenience. The portable system is probably cheaper.

Constant-flow oxygen

Below 20,000 ft you will probably elect to go with a *constant-flow system* and breathe through a *nasal cannula*. Somewhere between your nose and the tank is a flow regulator. You adjust it according to manufacturer's directions to get a steady flow of oxygen through the hose. Typically, either the instructions include a chart to tell you the flow rate for each altitude, or the altitude itself is printed on the flowmeter. The cannula directs that flow into your nostrils, where it dilutes with the surrounding air, but still provides adequate extra oxygen to you. At altitudes above 20,000 ft you will probably change over from a cannula to an *oxygen mask* plugged into the same system. Some masks have built-in microphones, which is nice because otherwise you have to take off the mask to be understood when you talk. Breathing pure oxygen at 40,000 ft provides about the equivalent oxygen supply to breathing normal air at 10,000 ft. During World War II, flight crews were in fact going up to those kinds of altitudes with this kind of oxygen system, but it was a bad idea. Dilution with outside air, mask leaks, fatigue, stress, and many other factors erode the performance of these systems badly. Most manufacturers of such systems suggest 25,000 ft as a practical upper limit for safe use. All our gliders in Telluride made climbs to 18,000 ft and used these systems. Most high-performance light aircraft have this kind of system.

Diluter demand oxygen

Above 25,000 ft you would be safer using a *diluter demand system*. This is a pressurized breathing mask system. Below 30,000 ft unpressurized air is supplied through the mask with an increasing amount of oxygen added. At that altitude, the proportion of oxygen is increased until it reaches 100 percent. Above 40,000 ft the airflow is pressurized. Thereafter, the pressurized pure oxygen could be used up to 45,000 ft. A mask tight enough to provide you with a pressurized airflow is very uncomfortable, and a pressurized helmet is an easier way to provide this kind of oxygen flow. Pressurized breathing is extremely uncomfortable no matter how it is done. You have to work very hard to force your exhalations after pure oxygen was forced down your throat. It is also very dangerous. You are at real risk of an embolism if you do this for any length of time.

Quick, name some small general aviation aircraft that can climb to 45,000 ft and has no pressurized cabin. I can't think of one either. Who uses this system? The military, whose aircraft could be subject to battle damage at those altitudes and depressurize. Also certain gliders attempting altitude records use these systems. Paul Bickle, setting the glider world altitude record in 1962, could have gone a few thousand feet higher, except that he felt he could no longer keep up with the demands of forced breathing required to exhale.

At 63,000 ft, something very bad happens. At that reduced pressure, water boils at 98.6°F, which is your body temperature. All the fluids in your system, including your blood, begin to boil. The only way to survive this is with a pressure suit or a pressurized cabin.

I have gone on at great length on this subject because so many pilots discount its effects. Portable oxygen systems are relatively inexpensive, widely available, and easy to use. Almost any airport can refill your bottle. Have this done only at the airport, by the way. Some medical oxygen has moisture added to be less irritating to your breathing passages, although today most hospitals use pure oxygen and add the moisture at the hospital. That moisture can freeze and block the lines. Welder's oxygen can have contaminates.

Dehydration

A longer-term effect of exposure to high altitude is an increased loss of fluids and a concurrent need for water. You have to drink much more water than at low altitudes. On a quick climb up to 15,000 ft followed by a descent back down within an hour or so, this might be no problem. However, if you fly up to a mountain airport at over 7000 ft and spend the night, you will adjust to everything much more easily and feel infinitely more comfortable if you drink copious amounts of water for your first few days. As with hypothermia, dehydration will increase your susceptibility to hypoxia.

This advice runs contrary to both the advice and the practice of many commercial pilots who deliberately avoid drinking before a long flight so they won't need to urinate. I advocate staying well hydrated, regardless of the inconvenience it might cause.

Fatigue

The reduced supply of available oxygen at higher altitudes gradually wears you down until you are well acclimated. If you have been up at altitude for less than 3 days, you are probably feeling fatigued.

Simply being tired can have a very pronounced effect on human performance. A very tired person will exhibit most of the detrimental effects of hypoxia. One of the very best student pilots I have ever recommended for a private pilot check ride had the very worst oral exam and check flight any of my students ever had. Why?

This particular pilot is one of those people who gives a lot more to others than he expects to receive. He volunteers many hours to the local sheriff's department for search and rescue. Specifically, the night before his check ride, five people were lost deep in the woods near timberline in a thunderstorm with inadequate clothing. He was out first searching for them and then rescuing them from 8:00 p.m. until 4:00 a.m. Next he was so cranked up on adrenaline from the rescue that he got no sleep whatsoever until 7:00 a.m. when he left for his check ride. Not just his motor skills and reactions, but also his memory and judgment were terrible. (And so was mine for letting him go ahead with the check ride.) Two days later everything was back to normal and his performance was superb. Did both of us and the examiner learn from this experience? We hope so.

Hypothermia

The heaters of most small aircraft are not particularly powerful. The insulation on most small aircraft does a very marginal job at best. The door seals and window seals are less than airtight. All this is by design. These systems are made as lightweight as possible for the sake of aircraft performance.

In the mountains, in the winter, you might be taking off in below-zero temperatures and flying in temperatures as low as −40°F, unless you fly in interior Alaska, where −60°F or more can be encountered. On those days, the cabin temperature might be cold enough to require bulky gloves and big insulated boots, a thick down parka, and heavy insulated pants. If you don't dress that way, the effects of cold will gradually overwhelm you. These effects include impairment of all motor functions, slowed thought processes, eventually uncontrollable shivering, and unconsciousness. *You will also be particularly vulnerable to hypoxia.* You can prevent this by dressing very warmly in layers and unzipping if you are too warm.

Hypothermia will be mentioned again on the next page in the discussion of cold-weather operations.

Cumulative effects

This is a good time to remember that the similar detrimental effects of fatigue, dehydration, hypothermia, stress, chemical poisoning, and hypoxia do not merely add up, although that would be bad enough. Instead, the effects of each make you significantly more prone to the effects of all the others. On a day when any one of the other factors is present to only a mild degree, the addition of high altitude can result in very diminished skills *and* judgment.

Cold-weather operations

While summer temperatures in the mountains can reach into the 90s, winter temperatures can drop well below zero. High altitudes in the middle latitudes and cold weather are inseparable. High altitudes anywhere in the world can see cold weather. Mauna Loa and Mauna Kea volcanoes in Hawaii, both just over 12,600 ft msl, see winter snowfall and a few determined skiers. Mount Kilamanjaro on the equator in Kenya have permanent snow fields although the effects of global warming in the 1990s are melting them. Mount Kenya still has glaciers.

Cold weather can be both a blessing and a curse for pilots. The considerations for cold-weather operations are summarized here. For convenience, they are broken down into temperature ranges.

Remember that the thinner air of higher elevations will aggravate most of the problems of cold-weather operations.

Above freezing

It is almost, but not quite, safe to say that there are no negatives and one big plus to this temperature realm. First, the good news. Colder air is denser. That means the density altitude is lower. The airplane flies better. There is more air available for the engine, so you get more power. The colder, denser air expands more with combustion in the engine, adding still more power, and the cooler air cools the engine better. Cooler, denser air is more stable and therefore less turbulent. The lower angle of the sun heats the surface of the ground less, producing less convection and with it less turbulence. A final little bonus is that most of the insects have been killed by night frosts or at least aren't flying, so your windshield stays cleaner. What could you possibly find to complain about in these conditions? Read the next paragraph.

Carburetor ice.   When the air temperature is anywhere from freezing up to 50°F and there is moisture in the air, ice can develop in the carburetor. If this ice builds up too much, it will restrict airflow into the engine, resulting in a loss of power, if not complete engine failure. Remember, air is cooled when it expands. At full power there is little risk of carburetor ice. But with reduced power settings, the throttle butterfly valve in the carburetor reduces the size of the air intake. After the air gets by the reduction, it expands and cools. Frost forms in the throat of the carburetor. This can happen when you are taxiing out for takeoff. It is most likely to occur when you throttle back to descend to land. Fortunately the aircraft manufacturer has provided you with a fix: carburetor heat. Pull the little knob next to the throttle, and heated air is directed into the air intake to melt the ice. It is good practice to use carburetor heat on any descent to a landing rather than try to calculate whether it is needed.

Airframe ice in flight.   Airframe ice does not form in above-freezing temperatures, but if the temperature on the surface is near freezing, temperatures not very far aloft are certainly below freezing. Flight into clouds or precipitation at these temperatures is likely to produce airframe icing.

From freezing to 20°F

In addition to the two problems mentioned above, we can add a couple more.

Airframe ice on the ground. Snow falling on an aircraft tied down outdoors can build up on the wings and control surfaces. If temperatures were warm when the snow was falling or if things warmed up during the day and then got colder at night, the airframe may have considerable ice buildup strongly adhering everywhere. A moderate amount of ice or snow on the fuselage is not necessarily a major problem. Ice buildup on the pitot tube and/or static port can render your airspeed indicator, altimeter, and rate of climb indicator inoperative or, worse, very inaccurate.

The biggest problem, though, is ice or snow buildup on the wings and flying surfaces. A wing develops lift because the air traveling over the top of the wing travels farther and has to go faster to do so. Faster-moving air generates lower pressure, which produces lift. The horizontal stabilizer and elevator are essentially a downward lifting wing balancing the weight of the nose. A stall occurs when the angle of attack of the wing gets to be too high and the airflow separates from the top of the wing. For most small aircraft, this occurs at

Airframe ice. Using wing covers saves an hour's work getting ice off the wings each morning before flight. Shelby Evans photo.

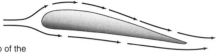

Smooth airflow over a wing. The air over the top of the
wing travels farther and faster. The lower pressure
produces lift.

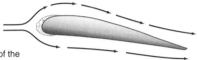

Light rime ice forming right on the impact point of the
leading edge only disrupts the smooth airflow slightly.

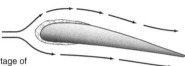

Smooth or clear ice forms over a greater percentage of
the wing's surface. It can build up very quickly and is hard
to lose.

The ram's horn configuration of mixed ice after a descent
into warmer air severely disrupts the airflow over the top
of the wing, causing a stall.

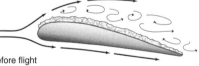

Snow or ice on top of the wing not cleaned off before flight
disrupts the airflow over the entire top of the wing,
destroying lift.

*Wing icing. Notice that snow and ice that accumulated overnight
when the aircraft is tied down outside can be potentially more
dangerous than ice accumulated in flight.*

an angle of attack of about 17°. A very rough wing top surface, particularly in the first 1 ft to 18 in back from the leading edge, can also
cause airflow separation and loss of lift. In this case the wing will
stall at a much lower angle of attack and at a higher speed.

Let's review an accident at Glenwood Springs, Colorado, several years
ago: The Cessna 177 Cardinal is loved by some pilots and hated by

others. While cruise is faster than that of the very similar-sized and powered 172, the low-speed flight characteristics are less forgiving. A charter pilot was anxious to get his two passengers down to Lake Powell. A very heavy frost still covered the wings. Because he was in such a hurry to get going, the pilot rationalized that he had flown many times with much more ice. Perhaps he had, but perhaps this ice may have been rougher-textured. In the event, the aircraft became airborne only slightly farther down the runway than it normally might have, but try as he might, the pilot could not climb out of ground effect. The south end of the runway in Glenwood drops off into a narrow little river canyon. The airplane essentially flew in ground effect off the end of the runway and settled over the edge of the cliff into the river. Fortunately for everyone, river levels are low in January. The plane came to rest upright with water just lapping over the door sills.

Snow on the wings.   So presumably you have elected not to take off with a load of ice on your wings. But a little loose snow would be fine. Brush it around a little bit. Yes, a quick check shows that this particular snow on your wings is the nice, dry, light kind that will blow right off as soon as you get moving. This was exactly the conclusion a freight hauler came to one afternoon in the early 1980s when his company called to get him to reposition his aircraft from Aspen down the valley to Rifle, Colorado, before an advancing storm hit. As with any story, there are at least two versions. One is that the plane was in fact clean, but after takeoff it simply encountered far more ice than it could carry. Quite possibly that is true. That is the version of the story the pilot told me, and I have no reason to ever doubt his word.

But the version of the story relevant to this chapter of this book is that the airport manager watched the aircraft taxi out with a considerable amount (in excess of 2 in) of snow on the wings. On takeoff it didn't all blow off as expected, and the aircraft couldn't climb. As it encountered just a tiny bit of icing in flight, it settled back to the ground in the Holland Hills subdivision of Basalt, Colorado, 20 mi away. A reasonable conjecture is that the remainder of the loose snow might have eventually blown off had the aircraft not encountered the mild ice later, and that had the wings been completely clean, the mild ice would not have had much effect. In retrospect, it would not have been all that much of a hassle to thoroughly sweep the wings with a push broom before flight. Most mountain airports have a couple of brooms available for that purpose.

Are you interested in performing a similar experiment with your plane? If not, you will have to go to whatever effort is necessary to clear all the ice and snow off the wings and flight surfaces before you take off. If the sun is shining, brisk sweeping will get the job done eventually. If the frost is light, just polishing it smooth with a towel will work. The best thing to have done would have been to put on wing covers the night before. Better yet, don't you wish you had a heated hangar?

Slow engine warm-up. Down to a temperature of about 20°F, a well-maintained engine with multigrade oil should not be too difficult to start. Warming up the engine will take longer than normal, not just because the engine is cold, but also because there is about a quart of cold oil sitting in the oil cooler remote from the engine proper. That oil in the cooler will not begin to circulate and warm up until the engine itself has warmed enough to open the thermostat. Until the engine has warmed up to the green arc on the temperature gauge, the colder engine parts will be fitting too close together, increasing wear, and the cold oil will not be doing an adequate job of lubricating, also increasing wear. The thick oil will be at too high pressure and may damage or even blow oil seals. Be patient! Warm-up will only add 5 min or so to your engine time.

Moisture in aircraft systems. During the time the engine compartment and cabin are above freezing, either moisture may condense in various places or rain or snow may blow into things. Overnight, that moisture may freeze and cause a number of either minor or serious troubles.

The pitot and static ports may be frozen over, which will render the airspeed indicator and altimeter inoperative. Moisture frozen inside of control cables can jam them. You could find a completely frozen throttle, mixture, or pitch control. Less serious but equally jammed, your cabin heat and vent controls are also cables.

Any instrument or radio could have enough internal frost to render it inoperative.

Moisture frozen on the starter can prevent the starter Bendix drive from engaging the teeth on the flywheel. The engine won't start. Go back inside and warm up, or spray the manufacturer's recommended silicon lubricant on the starter shaft when the engine is warm to prevent a recurrence.

A thorough preheat will miraculously cure these myriad problems. While the engine is running and the cabin heat on, they will generally remain cured except in certain circumstances. These circumstances are a rapid climb to very high altitudes where the temperature is below zero and wet things begin refreezing at whatever setting they are currently in. A descent to warmer altitude will again free everything up. Prolonged normal operation at temperatures where everything works will eventually get rid of most of the moisture, and the problems will be gone. Naturally more moisture in the system will cause them to recur.

Frozen landing gear. When you taxi and take off in slush (or possibly even in standing water), slush and ice will build up on the landing gear and in the gear wells. After you retract the gear and climb to colder altitudes, the gear can be frozen in the retracted position. If you descend to warmer elevations before you lower it again, you may never realize you had a problem. If it is still cold when you land, the gear may be frozen in the up position. To avoid the problem, it is sometimes advisable to cycle the gear a couple of times on the climb in order to break out the ice as it forms.

Very cold hydraulic fluid in the gear retraction mechanism can be a problem as well. It can require so much pressure from the gear pump to move the gear that a circuit breaker will trip and the gear cannot be lowered. The solution is to remember that there is a manual gear pump lever somewhere (usually between the pilot seats). You can also reset the circuit breaker and try again.

Temperatures below 20°F

Oil this cold does not provide adequate lubrication. Even multigrade oil such as the 15-50 weight you probably use will be as thick as honey, which means that the engine starter motor may not be able to crank the engine at all. Certainly your battery isn't helping much. At 32°F, your battery with full charge produces only 40 percent of its normal cranking power. This is a blessing because it means you are going to be forced to warm the engine before you can start it. You can do that by waiting until later in the day, getting the airport to preheat the engine with a portable propane heater, or plugging in an electric engine heater which is permanently mounted to the bottom of the oil pan. While the propane preheater will warm everything under the cowling, the electric heater will not warm oil in the oil cooler. In either case, after the engine starts, you still have to warm it up to normal temperatures before you take off.

Single-weight oil, such as you would use to break in a new engine, is a nightmare in the cold. At 20°F, it has practically congealed and is worthless for lubrication. You might need this kind of oil to break in a new cylinder—opinions differ; but otherwise you will be very glad you changed it before cold weather.

Cold has an effect on all your instruments as well. Until they are warmed up, they won't function properly. Gyros will not spin fast enough, and moving parts will stick, so the heading indicator and attitude indicator won't work. Modern electronic instruments and radios will work better than the older radios still found in some small planes, but won't work anywhere near as well as they should until cabin temperatures get above freezing.

Water droplets in the fuel are now ice droplets and can block fuel lines. Still liquid water in the fuel will freeze in the carburetor and block it.

Temperatures below 0°F

The problems are the same as with temperatures below 20°F, only more so. In particular, your engine oil has congealed into something like very bad-tasting Jell-O. Single-weight oil may resemble something more like Vaseline jelly, if not candle wax. It is absolutely impossible to start the engine when it is this cold. If you somehow managed, some parts of it would heat up much faster than others, and a possible result would be a cracked engine block or cylinder. Preheating is absolutely mandatory.

Temperatures lower than −30°F

Fortunately for everyone, these temperatures are seldom encountered on the surface anywhere except in the interior arctic. At temperatures colder than −40 to −50°F, most aluminum alloys lose considerable strength. Normally, if you encounter these temperatures in flight, the air is very likely very smooth and no problems ensue. The stresses of a normal takeoff or landing could at least theoretically cause some cracking of the airframe in these very extreme temperatures.

The drill for starting an engine that has to sit outside in these temperatures is to drain all the oil into a bucket when the engine is still warm. The next morning, the oil is carefully heated indoors on a stove and then poured back into the engine. Meanwhile, the engine was being preheated, and you start it before everything can freeze again.

Messing around doing all this will probably give your fingers frostbite when you take off your gloves to do preflight checks, and you may freeze your fingers to various metal parts when you touch them. And don't touch your tongue to the propeller! You have to want to go flying really badly to go through this process. At least in the arctic, many people do go through this process, not because they enjoy flying, but rather because small aircraft are the only link to the outside world and the flights can be a life-saving necessity.

Reminder about hypothermia

We touched on hypothermia in conjunction with hypoxia. Let's have a quick review.

Cold, not merely extreme cold, has pronounced detrimental effects on pilots as well as airplanes. Without going into unnecessary detail, it should be enough to remind you that if you are not dressed warmly enough, your judgment, coordination, and motor skills will all begin to deteriorate. Rapid shivering could seriously compromise your ability to fly the airplane. Small aircraft cabin heaters are only marginally adequate at best, cabin insulation is poor, and cold air always seems to be blowing in everywhere.

Here are a good example and a bad example. Several years back my son Leif and I found ourselves flying home to Glenwood Springs from Denver in an older Cessna with a very marginal heater. Denver temperatures on the ground were a very unusual −20°F. Temperatures at altitude in flight were showing −40°F on the outside air temperature gauge. While we were less than absolutely comfortable, we *were* prepared. Both of us were wearing huge bulky gloves, thick down parkas, and warm-up pants. Besides that we were actually flying in our sleeping bags with just our shoulders sticking out and the foot of my bag unzipped just enough to use the rudder. We had no problems and the flight was uneventful. The heater was repaired before the next flight.

In the fall of 2000, photographer Ron Kanter and I had been trying for weeks to get just the right photograph of the town of Telluride with all the aspens at the height of their fall colors. Suddenly when both of us were at work at the airport, the moment arrived. Quickly we removed one door of the airplane and took off. Ground temperatures were about 60°F, and in the bright sun we were both wearing shorts and T-shirts. At 16,000 ft temperatures were close to freezing if you figure in the wind chill from the missing cabin door. Now 32°F hardly counts as an extreme temperature, but we were not prepared for it. In a very short time Ron's fingers were almost too stiff to work the shutter on the camera, much less load film. My lips were too stiff to speak properly on the radio. The cannula froze to my face. Both of us were experiencing violent uncontrollable shivering. But if we stayed up just a few minutes more, we would get the shot!

We had stayed up too long. On the way down I was definitely too stiff and too cold to fly at my very best. Ron's hands, which had been out in the airstream, were just beginning to show signs of frostbite. A longer exposure could have been the start of a chain of events leading to more serious troubles than mere discomfort. Rewarming after we landed was a long, drawn-out, and painful process.

We did get the shot. You can buy the poster today at the airport desk.

Cold-weather accident precautions

In January 2002, an airline pilot trainee rented a small aircraft at Farmington, New Mexico, and flew into the mountains between Durango and Telluride for some sightseeing. Conditions were not extreme—I was flying a lesson with a student in the same type of aircraft at the same exact time not 2 mi away. But by violating some basic rules of mountain flying, the pilot managed to get trapped at very low altitude in a narrow valley with no room to turn back. He saved his passengers' lives by executing a deliberate stall landing into the treetops. The pilot was not badly injured, but both passengers were. *He* hiked out to get help. *They* spent the night at 10,000 ft in January in the mountains in near-zero temperatures *wearing only shorts and T-shirts!*

They survived because they could start a fire. Check your pockets right now. Do you have any matches? I am checking. I don't. Are there matches in the plane? I think I have two books of them in the first aid kit, but maybe I should check that before the next flight.

And the story does not end there. The rescuing Air Force Special Operations MH 53J Pavelow helicopter carrying $40 million worth of secret avionics, sensors, peculiar gizmos with little glowing lights, and so forth hit some treetops with its rotors as it hovered in the dark for the rescue, and the Cessna pilot was in his second crash in one day! A second rescue helicopter arrived the next morning, but at least by then everyone had warm clothes.

A successful forced landing in extreme cold can put the airplane's occupants in an unsurvivable situation unless they have taken some precautions. It is absolutely necessary to carry clothing warm enough to be comfortable outdoors in the temperatures you would encounter. Carrying sleeping bags should be almost mandatory. You should also be able to make a fire and have enough food to spend the night. Heavy snow boots would allow you to walk out. Searchers will be as hampered by the cold and bad weather as you will, so rescues will be prolonged. *Every winter there are plenty of stories like this, and far too many stories of people surviving the forced landing only to die from cold-weather exposure afterward.*

In the normal course of events, the pilot in the previous example will have completed his airline training program by mid-summer 2002, and he will begin flying right seat in a Dehaviland Dash 8 as first officer. Do you think this taught him anything? Are you wondering how to tell if you are a passenger on one of his flights? He is the guy wearing long pants and a long-sleeve shirt.

2

Mountain weather

We generally discuss the weather as though it were either a constant or at best predictable phenomenon. This of course is absolutely false. No where on earth is this less true than in the mountains. Recall once again the fact that one-half of the earth's atmosphere and most of its weather occur under 18,000 ft. No less than 92 percent of the moisture in the atmosphere is found under 18,000 ft. A range of mountains rising up over 14,000 ft effectively disrupts at least 50 percent of the airflow across the continent and potentially dams up 80 percent of the weather. Something is always going to happen which could potentially change all the weather as humans on the surface perceive it *whenever* an air mass meets the mountains.

A storm system may be lifted in altitude; all its moisture may condense out and smother the mountains with snow. The storm system may park right over you for a week. With little or nothing predicted at all, you can get a foot of snow. Then again, as a storm is lifted, high winds aloft may blow it right out of the area. Perhaps the blockade of the mountain front may deflect it away from you entirely. Frequently a high-pressure system will appear to be dammed up against the west side of the Rocky Mountains for weeks on end, producing clear sky, unlimited visibility, no turbulence, and plenty of opportunities to fly while every ski area in the region will be praying for snow. Then again something altogether strange may occur.

Back in the late 1980s, weatherman Woody Woodrow in Jackson Hole, Wyoming, observed perhaps the largest and most intense winter storm system he had ever seen move slowly out of the Aleutian Islands toward the northwest coast of the United States. At that point it was of primarily theoretical interest only. But as days went by, the storm track began to stabilize on a heading directly toward him. With each day, more factors appeared that strengthened that prediction. At the same time, the storm itself showed no signs of abating.

The mountains of Washington and Oregon were hammered. Roads closed. Avalanches isolated small communities. Woody began to call the storm The Big Kahuna!

In addition to reporting the weather locally in Wyoming, Woody Woodrow is one of the premier avalanche forecasters in the world, a leading authority on mountain weather who gives seminars for ski patrols and other outdoor professionals every year. So when Woody began predicting The Big Kahuna, people began stocking up on extra food and emergency supplies. And as The Big Kahuna began sweeping toward them across southern Idaho, closing schools and blocking the interstate highways, even the few remaining unbelievers put chains on their cars and waxed their powder skis. By late afternoon, from the top of the gondola at the Jackson Hole ski area, you could actually look out to the west and see the killer storm approaching.

But the next morning, the skies were hazy and clearing. There was about half an inch of snow on the ground. That was The Big Kahuna? That was it. The monster storm had smashed up against the west face of the Teton Mountains and just faded away. Why? How? Who knows? And, by the way, Woody Woodrow is still one of the absolute best avalanche forecasters in the business. He himself tells this cautionary tale to his seminar audiences.

Weather on a continental scale can be observed by satellite in real time. Major trends can be seen, and large-scale weather patterns are increasingly accurately predicted. You can watch it on television.

But this is not so in the mountains. Mountain weather can be seen in real time on radar, but not predicted well beyond a couple of hours. Even then, predictions vary in accuracy. Long periods of high pressure and clear skies can be predicted over the mountains with about 90 percent accuracy. The behavior of storms, the movement of low-pressure centers, of moist air masses, and even of winds aloft is not quite so exact a science. In these conditions, 3-day forecasts have an accuracy of about 20 percent. Once you get into the mountains, there are simply far too many factors at work influencing the weather to be understood and analyzed.

That is not to say you should ignore the 3-day forecast. On the contrary, if you plan to fly regularly in the mountains, you will probably become an avid follower. But you should not be surprised to see the forecast revised frequently.

Global weather

Global weather is the result of solar heating of the earth's surface, evaporation and precipitation of moisture, and rotation of the earth. Global weather can be very accurately modeled for a perfectly spherical earth. Unfortunately, or fortunately, depending on your philosophy, the earth is not perfectly smooth. Two-thirds of it is covered with water, one-third with land. Mountains stick up halfway through the atmosphere. Truly accurate long-term weather prediction is not yet possible, given the myriad factors at work. Short-term local forecasting has dramatically improved in recent decades, due in no small part to the vastly increased use of satellite monitoring.

We have now put into play enough elements to produce a very complicated and changing weather map. Nevertheless, for any region of the globe we have recurring and relatively predictable seasonal weather patterns—at least we have that from a computer modeling standpoint. In reality, we have yet to allow for the mountains. Will the continental-scale air masses go over or around the mountains? In so doing, will they be slowed or accelerated? Will they lose their moisture or not? As anyone who lives in the mountains can tell you, any of these things might occur. The results will be slightly different each time.

There are two catchy phrases that everyone living in the mountains everywhere in the world hears every day of their lives: "If you don't like the weather, wait 15 minutes." "Only God and newcomers can predict the weather." While this is all too true, there is still some degree of reasonableness to large-scale forecasting. Winter in the arctic is typically very stabile and often dry. In the mountains of North America, there is a pattern of mild summers with afternoon convective storm buildup, and cold winters with low-pressure storm systems alternating with periods of high pressure and stabile air masses. The jet stream will alternately loop south of low-pressure systems, pulling down cold arctic air, and loop north of warmer, high-pressure systems.

You might not know what the weather will be doing next week. But in a general way, you do know what it will be doing next August. But what if your plans include a flight to Sun Valley, Idaho, on Thursday?

Weather predictions and current information sources

There are people who swear by the *Farmer's Almanac* or the swelling of their bunions, and I am not going to argue with them. Your feet have no bunions? Neither do mine. Luckily, there are other equally valid sources of weather information at our disposal. Rather than attempt to sort through the near infinite mass of weather data available, you can turn to professional weather services.

Three-day forecasts over mountainous terrain are among the least accurate weather predictions available. Predictions for weather events occurring within the next 3 h, about the duration of a flight, are more accurate, and their accuracy has been greatly improved in recent years by improved reporting, but hampered by the complexity of the pattern over the mountains. This might be useful from a planning standpoint, but expect predictions to change as the date draws closer.

Weather information becomes critical when you are about to depart on a flight. That is, critical weather information for you involves weather over the next 2 h. Yes, the accuracy of these predictions is better, in part because a lot of it is based on very current observations. Remember, though, that in the mountains—in any mountains anywhere in the world—there is a scarcity of official weather reporting stations, and not all your route will have current reported weather. Remember also that weather over the mountains can change dramatically in a very short time, periods of time far shorter than the duration of even relatively short flights. The official weather sources can tell you whether the weather is improving or deteriorating. If it is deteriorating, they really cannot say how quickly with a high degree of certainty. Do you really need to fly that day? If it is improving, then you can very reasonably take off into conditions which are acceptable at your point of departure, but uncertain over your route.

Some sources of weather information are as follows:

The Internet

The most popular of dozens of Internet weather services is called DUATS (www.duats.com). Another is www.weathertap.com. Some people are so engrossed with the Internet that they quite literally forgo the rest of their lives. Other people begin to visibly salivate and foam at the mouth in anger. If you fall somewhere in between, it is possible to subscribe to a variety of weather services, some specifically devoted to aviation. You can bring up current and predicted weather in text or map form as well as current satellite and radar imagery. If you live near a busy airport, it is likely that the FBO subscribes to a satellite weather service and has a computer dedicated to bringing up this same information. (See below.)

The Weather Channel

This is perhaps the most watched cable TV station in the United States! As with any weather service, the farther into the future you get, the less reliable the prediction. But what you do get is real-time satellite imagery combined with real-time radar and the best meteorologists in the business. You don't just hear the prediction; you see images of the weather and get a reasonable explanation of what's happening. You know you are a real pilot when the Weather Channel is your favorite source of entertainment.

Airport weather computers

These days all large and many small airports have a computer in the FBO devoted solely to weather information. While there are a couple of different services, their products are essentially similar. The computer has a direct satellite feed. With a click of the mouse you can get a wide variety of radar and satellite images; isobar charts; graphic displays of current and predicted weather including IFR, icing, and turbulence airmets and sigmets; and multiple pages of text. The beauty of these systems is that they are at the airport and are direct feed rather than delayed information.

All the above varieties of weather services ultimately derive their information from the same sources, which are satellite imagery, surface radar, and ground station reports. To varying degrees, they are using predictions from the National Weather Service as well as their own prediction capabilities. The fact that predictions from these sources may disagree is a reflection on the unpredictability of weather rather than on the superiority of one service over another.

Flight service

This service is free (not exactly free—it is funded by the FAA from aviation fuel taxes) and easy to access over the telephone. Anywhere in the United States dial 1-800-WX BRIEF. You get a menu of recorded weather information about your region, or you can talk to a real live weather briefer. The briefers have at their fingertips a vast amount of information concerning both current and predicted weather all along your route of flight, at all nearby airports and at your destination. They are very well trained and knowledgeable. That allows them to interpret the same information you might be able to obtain from other sources better than you could do it. Their experience with their own regions of the country helps them make reasonable guesses about which of many seemingly conflicting bits of information is likely to be accurate and why. Computerized weather comes up as a set of strange and sometimes indecipherable abbreviations, acronyms, secretly coded numbers, TAFS, METARS, and so on. Live briefers who decipher that kind of thing all day long are more likely than you to get it right.

You call them up, push whatever the recorded message tells you to push for a briefer, and then say. "My name is Charles Lindbergh, my number is NX211, a Ryan monoplane, and I need a standard briefing for VRF flight from New York to Paris departing 7:00 a.m. tomorrow local time." Or perhaps, "My name is Wright, I am a glider pilot inquiring about surface winds this afternoon at Kitty Hawk, South Carolina." They need your aircraft registration number to enter in their database before they can give you the briefing. Giving them your name is just being polite.

En route weather information

It is important to remember that Flight Service can also provide up-to-the-minute *en route* weather advisories *when you are flying*. Anywhere in the country, a call on radio frequency 122.0 gets you *Flight Watch*. These are the same people as 1-800-WX BRIEF.

Are the clouds building up rapidly in front of you? Is the turbulence getting ugly? Just try this procedure: "Denver (Salt Lake, etc.) Flight Watch, Cessna 1234 near (city, VOR, etc.)"

You make their job easier if you tell them where you are because the person on the other end of this radio conversation is simultaneously monitoring several remote communication outlets on this same frequency. But if all you say is, "Flight Watch, Cessna 1234," someone will get back to you. *Wait for them a moment or two* because the person you want to talk to is monitoring several stations and may be talking to someone else whom you can't hear. When they get back to you, you say, "Denver Radio, Cessna 1234 is over X en route to Y.

Do you have current weather at Y?" or "The weather ahead looks pretty bad. What is the weather like south of my route? Do you have reports of cloud tops? Is anyone reporting turbulence?" Or ask for whatever weather information you want. You are talking to a briefer with all the available weather information right in front of her or him. One of the most useful and current bits of information this person has is pilot reports.

PIREPS. Once again, please forgive the government's insistence on putting everything in secret codes. They can't help themselves. *PIREPS* are pilot reports. The briefer either over the phone or on the radio at Flight Watch can tell you something, such as "Cloud tops are reported at 16,000 ft, bases reported at 12,000 ft. We just had a report of occasional moderate turbulence at 18,000 ft by a Beech 1900 ten miles east of the Red Table VOR."

It doesn't get any better than that. Another pilot just like you flying somewhere ahead of you along your intended route has told the folks at Flight Watch exactly what he or she actually encountered. You could be the one informing them of the weather as well. Call 122.0 on the radio, and report any sort of weather you encounter that is different from what you expected. You give them your name, type of aircraft, and what you are observing.

AWOS. AWOS is the acronym for automated weather observations. You usually think of this as landing information at an airport, because many, but by no means all, nontower airports have AWOS. A computerized voice plays continuously, saying, "(airport name) Automated Weather Observation (sometimes Zulu) wind (speed and direction) peak gust (speed) visibility (distance in miles) sky conditions (clear below altitude or few clouds at or ceiling in feet or whatever) temperature (in Celsius) dew point (in Celsius) altimeter setting and remarks (density altitude, for example).

The AWOS at my home airport in Telluride will follow this critical information with a lengthy sermon about noise abatement procedures involving landmarks that visiting pilots don't know about just when you were hoping for a wind update on final. Then there is a silent pause just long enough to convince you that your radio is on the wrong frequency.

AWOS frequencies are found next to the airport on your sectional chart and in publications such as *Flight Guide* and the *Airport Facilities*

Directory. In *Flight Guide, you will often find a telephone number for AWOS as well.* The normal use of AWOS is to figure out which runway to use and to reset your altimeter before landing.

Listening to AWOS en route will give you updated altimeter settings and current surface winds. Also the difference in altimeter settings between airports on either side of a ridge may give you a heads up to expect airflow from the high-pressure side to the low.

In the mountains, AWOS located at airports really only gives you weather information for the valley floors where the airports are located but leaves you deep in the realm of speculation about weather up by the mountaintops and at the passes you will be flying over.

Colorado remote mountain AWOS. *Beginning in 2000, the Colorado Department of Aeronautics began installing remote AWOS at certain heavily traveled mountain passes.* These frequencies and locations are to be printed on the sectional charts. Knowing wind speed and direction over a mountain pass can contribute significantly to your safety. The first group installed was at Corona Pass, Vail Pass, Monarch Pass, Copper Mountain, and Wolf Creek Pass along the Continental Divide; La Veta Pass, which connects the San Luis valley with the plains; and Monument Hill between Denver and Colorado Springs. Flight Watch, on your radio at 122.0, has access to these reports over telephone lines as well.

On my regular weekly trips from Telluride to Denver I have begun to check the Monarch Pass AWOS, despite the fact that if I am at 18,000 ft, the surface weather at the pass does not directly influence my flight. The reports are frequently eye openers. During periods of very high atmospheric pressure, although the winds aloft might only be blowing 20 kn, they are frequently funneled through Monarch Pass at considerably higher velocity—occasionally twice the velocity aloft! A small plane climbing over the pass from the east would often encounter very strong turbulence and downdrafts on a day when nothing of the kind was predicted.

TWEB and HIWAS. A TWEB is a *transcribed weather broadcast,* that is, a tape recording. A HIWAS is a *hazardous in-flight weather advisory service,* which is also a recorded message. The legend on the sectional chart shows what to look for. Either of these may be found at a very-high-frequency omnidirectional radio range (VOR), and their presence is indicated by a tiny circle with a letter in it in the box of information for the VOR. Weather information is transmitted continuously over the VOR navigation frequency. Unfortunately for you, you are very unlikely to find either of these services located at any VOR in the mountains. Why not? VORs in the mountains are often located in remote areas, sometimes on top of high mountains. It is difficult to access the sites and service the equipment. Adding more services is expensive and impractical.

Summary.   There are several sources of weather available. At the extremes of good weather and of bad, they can give you a very accurate report to help you decide whether to fly. If you fly on a set schedule, conditions are often far from perfect, but not impossible either. You typically find that the predictions and reports are always somewhere in between and incomplete.

Decision making based on Flight Service reports in conditions which are neither impossible nor excellent is very hit or miss. As various mountain pilots have observed, "*mountain obscuration, VFR flight not recommended*" is almost always included in the briefing. If you took that at face value alone, you would never go at all. This would not really be a bad idea—it is never wrong to go only in optimal conditions. But if you tried it a few times, you would discover you could usually get away with it. Then you would begin to disregard the briefing advice and always go anyway, only to discover it was even worse than they said.

While Flight Service may sometimes know that most of the mountains are obscured, predictions of cloud tops are a little more art than science. If you call very early in the morning, there will not be any pilot reports of tops yet, although the briefer may be able to hazard a guess based on the weather pattern of the previous day. The mountains may well be completely obscured, but you would fly over them in clear weather above the relatively low cloud tops. Better than 75 percent of my winter flights from Telluride to Denver are done this way.

Flight Service personnel themselves are all aware of this and are gradually changing the wording to say "*areas* of mountain obscuration, VFR flight not recommended *in the areas of mountain obscuration.*" With this phrasing, they are not strongly advising you not to fly at all, they are alerting you to the possibility of trouble along the route. They are admitting incomplete knowledge on their part. Your local knowledge and the specifics of the route will help you to make your decision.

"Occasional moderate turbulence below 16,000 ft msl" is another phrase you hear during briefings so frequently that you begin to suspect it is a recording. Moderate turbulence in a small plane is pretty uncomfortable for most people. Moderate turbulence in the immediate vicinity of the rugged terrain of the mountains includes very strong downdrafts that you could not possibly climb through. Moderate turbulence near the ground can add unacceptable danger to a takeoff or a landing. So you aren't going, right?

But perhaps you could. Occasional moderate turbulence is a predicted condition. *Occasional* is a key word. If you are flying a plane comfortable with higher altitudes, you could (and if you fly on a schedule, frequently would) employ your mountain knowledge to find the safest and most comfortable route to get up through the turbulence. You cruise above it to your destination, and then you plan the best route to descend through it. Pilot reports available from Flight Service would tell you what other people were actually encountering.

Even in a very light, small plane which could not practically get above the predicted turbulence, you might often be able to select a route that avoids most of the problem. "Occasional moderate turbulence below 16,000 ft" is typically associated with high winds. The turbulence is the result of the wind blowing over the ridges. By avoiding the areas downwind from ridges, you also avoid most of the turbulence. Of course, not all destinations allow that choice of routing. For example, in strong westerly winds, the descent into and the climb back out of the Denver area are all right downwind from the Front Range of the Rocky Mountains. Somewhere on the climb or the descent you will almost certainly encounter turbulence. Your choice of route can minimize it, but not really eliminate it. Your mountain flight knowledge and experience are going to play a key role in your decision making.

It was worse than we expected, but we decided to press on.

With either mountain obscuration or turbulence predictions, you will frequently encounter conditions different from those predicted. You will sometimes make the call not to go on days that turn out to be fine. Don't do what I always do after that: Since flying is how I make my living, when I cancel a flight, I don't get paid, so I begin second-guessing myself almost from the instant I decide to cancel.

The bigger problem in terms of safety is obviously deciding to go, but then encountering conditions worse than you anticipated. The majority of small aircraft accidents everywhere, not just in the mountains, involve the pilot deciding to press on in deteriorating circumstances or even just when she or he knew something was not right. (Statistically, men are more prone to this error than women.) There is a certain mind set that we all fall into that magnifies the importance of getting to a destination beyond all reality, and in so doing it obscures the alternatives. It is a seductive mental trap. Your life will be so much less complicated once you get there. You only have to endure these terrible conditions another few minutes, and then everything will be better. It is always much easier to see what the right decision would have been after the fact, when you are discussing someone else. There are a lot of examples every year of someone else doing something dangerous. If you are willing to change your plans and turn back, at least it won't be everyone else discussing you, and your bad example won't find its way into a future edition of this book.

(Another interesting aspect of the system is that if you fly the same route all the time and call for the briefing very early in the morning, you will be talking to a very small number of Flight Service employees who are on duty during those hours. Sometimes they will recognize you and over time will have a more accurate view of your personal capabilities.)

Now having checked the weather in depth, you are about to fly. Or are you?

All the above information was incorrect!

It is only prudent to check all the weather information available to you before flight in the mountains. After all, you cannot see even as far as the next valley, and the weather there could be extremely different from that where you are now. But don't stake your life on the accuracy of the reports you just received!

Weather predicting over the mountains is seriously flawed. There are far too many variables at work in that environment to give predictions anything like the degree of accuracy they have in other environments. Weather reporting stations in the mountains are too far apart to give complete coverage of the entire area they serve.

While weather reporting stations are sometimes far apart, weather phenomena in the mountains can be extremely limited in area. Here are a few examples.

Although the San Juan Mountains in southwest Colorado might be obscured in snow and clouds, the weather at the Telluride Airport might be just fine. Condensation can occur when an air mass is lifted and cooled. The airport in Telluride is in a west-facing valley just inside the mountains. An air mass blowing in off the desert will be lifted as it hits the mountains, but the air coming up the valley hasn't been lifted quite yet and the airport may have a big blue hole over it, even though snow is falling 4 mi away at the ski area. Westerly winds packing air into the west-facing Telluride valley frequently produce an air pressure 0.5 in higher than that in nearby Montrose, Colorado. This translates to a pressure altitude 500 ft higher or 3°F warmer. Either could be enough to produce a nice blue hole of clear air right over the airport. More often, the weather at the airport might be marginal, but not impossible. Weather scarcely 5 mi west might be perfect.

Also at Telluride, a combination of not excessively strong south winds and high atmospheric pressure may cause the *local* winds to venturi through Lizard Head Pass, south of the airport. That strong, very localized wind is aimed directly at cliffs adjacent to the runway and will produce a strong crosswind and low-level turbulence localized in the vicinity of the airport, even though winds 5 mi to the west are quite mild. If you got a weather briefing for Colorado, things would sound acceptable, though not perfect. Only in a very light plane would you encounter any turbulence and most of that near the ridges. Only a pilot report from a very small plane landing at Telluride would alert you to turbulence on the approach.

The Gunnison, Colorado, and Eagle, Colorado, airports are situated in the bottoms of valleys drained to the west by narrow canyons. In effect, they almost sit at the bottoms of closed basins. In the winter during a cold, still night, cooling air flows downhill into the valley and fills it up like a bowl of soup. Residual moisture condenses out of the cold air. The entire state can be experiencing a crystal-clear blue sky day, but in the early morning hours, Eagle and Gunnison are IFR in a low layer of thick valley fog. If you call Flight Service to check the weather, they will tell you Gunnison is solid IFR with half-mile visibility, zero ceiling or indefinite ceiling in fog. But can you fly over the mountains to Denver? According to that report, no. But in fact, a report like that one in winter may well be an indicator of a nearly perfect day. The top of that thick fog is probably less than 1000 ft off the ground. The fog accumulated overnight because the air is very stabile and the night was very clear with no wind. The weather, except right in the bottom of that valley, is as nice as it ever gets.

In most mountain valleys, the wind is either blowing up the valley or blowing down the valley. Wind across the valley is extremely unlikely. For that reason most mountain airports are aligned with the direction of the valley, and the airports at Leadville and Buena Vista, Colorado, in the Arkansas River valley are no exception. Every so often, even though weather throughout the region is relatively mild, small-scale local conditions produce a flow of very cold air coming down off the Collegiate Mountains along the west side of the valley. Despite mild weather elsewhere, those two airports may experience a strong, cold 90° crosswind.

At any mountain airport in the Rocky Mountains, winds in late spring can be horrendous. Spring can combine the very high winds aloft of winter with the unstable air and high-temperature gradients of summer.

For many years, pilots attempting to cross Hagerman and Independence passes between the Roaring Fork valley near Aspen, Colorado, and the Arkansas valley near Leadville on the other side of the continental divide reported abruptly hitting very strong wind shear turbulence and downdrafts west of the passes on days when the prevailing west wind should have made that side smooth and lifting. Several of these incidents were responsible for actual crashes involving very experienced pilots. What was going on?

The mystery was finally solved (and many lives saved, perhaps including my own) when Margaret Lamb and Sue Baker discovered that on the dates of the crashes, the Leadville altimeter setting, east of the divide, was substantially higher than the Aspen altimeter setting, west of the divide—the exact opposite from what you would normally expect. Despite westerly winds aloft over the region, *there was a sometimes very strong wind blowing over the passes in the opposite direction* at altitudes of 1000 ft off the ground and below as the high-pressure valley drained air into the low-pressure side. The small aircraft climbed toward the pass with a tailwind and some lift. Just as they were about to cross, they flew into this low-level opposite-direction wind and were smacked hard with wind shear, sink, and turbulence. Since the study was published in 1986, various people have looked for and observed the same phenomenon over other mountain passes as well. Three fatal crashes over the Uncompahgre Plateau near Montrose, Colorado, in the period between 1993 and 1997 fit this kind of pattern. Would Flight Service have been able to warn you of this trap? Probably not, and aircraft flying high enough to contact them with pilot reports would have been well above this localized wind and never felt it. If you diligently checked the Aspen and the Leadville altimeter settings, you might be tipped off to watch out for the possibility of such a wind, but no one at Flight Service would have alerted you to do so, and even then, the reverse-direction wind might not always be present.

Observe the weather

So the weather you actually encounter when you are there is the only trustworthy source of weather information in the mountains. But what are you actually seeing and feeling? The ability to understand and react to or avoid a very small-scale weather phenomenon is a critical skill for a mountain pilot.

3

Reading mountain weather

Observing weather

It should be, but too often is not, self-evident that the weather you actually encounter is the "real" weather. The predictions and observations you dutifully noted from Chap. 2 are not. Yet it is always amazing to learn how many pilots have the attitude that having listened to a weather briefing, they are now covered. Any deviation from the predicted weather is someone else's fault.

It cannot be said too many times that you may have to change your plans, even to the extent of turning back completely, if you encounter weather worse than predicted. No matter how obvious this idea seems to be on paper, people in the air fail to heed it in large numbers every year. Dick Arnold offers an axiom for all aspects of mountain flight: Assume you will never get all the way to your destination, so plan alternates.

At the risk of preaching, it really doesn't matter whose fault anything is. If you are the pilot, you alone have to deal with whatever you encounter. If you are like most of us, then you are going to have to cultivate a willingness to turn back or not even depart in the first place at times when you are reasonably sure you could still get to your destination. This can be particularly difficult when you are carrying passengers with scheduling needs.

So we have finally found our way to the ultimate weather observer—you, looking out the windshield. But how can you make sense of what you are seeing? Almost the only visible weather is clouds.

Clouds

"I look at clouds from both sides now, from here and now, and still somehow, it's clouds' illusions I recall, I really don't know clouds at all," sings Judy Collins. Clouds are visible moisture in the air. Knowing

73

how the various cloud types form gives you valuable clues about air movement.

The air can hold a certain amount of moisture in suspension. The *saturation point* is the point at which the air can hold no more moisture, but the saturation point is temperature-related. Warmer air can hold more moisture than cooler air. The *dew point* is the temperature at which the *saturation point* is reached. When air cools below the dew point, excess moisture in the air condenses out, forming clouds (or rain or fog or snow). Air has reached 100 percent humidity when its temperature reaches the dew point. When that point is reached, clouds form.

Summarize this as follows: When warm air cools beyond a certain temperature, clouds form. This temperature will be different every day, depending on how much moisture is in the air and on the air pressure, but *when you see a cloud, you are seeing the result of cooling air. Cooling in turn usually has something to tell us about air movement.*

Since more explanation confuses rather than simplifies this fact, let's move on to specific cloud types.

Cloud types

Cumulus-form clouds

Cumulus clouds vary in form from something like clumps of cauliflower to giant towering thunderheads. Their overdeveloped states include thunderstorms, hailstorms, and tornadoes. Fundamentally, cumulus clouds represent rising or lifted air.

As you go up in altitude, air pressure is reduced and air temperature is lowered. Unsaturated, dry air cools at a rate of about $5^{1}/_{2}°F$ per 1000 ft in elevation. This is called the *dry adiabatic lapse rate.* All else being equal, if the dry air temperature at sea level is 60°F, the dry air temperature at 10,000 ft will be 5° F. The standard lapse rate for air of normal humidity is around $3^{1}/_{2}°F$ per 1000 ft, so on the same 60°F sea-level day, air of normal humidity would have a temperature of 25°F at 10,000 ft. To look at this in a different way, if the air temperature on the ground in Denver, Colorado, were comfortable shirtsleeve weather, the air temperature at the top of Loveland Pass 25 mi to the west would be below freezing.

On any given day, the actual rate at which air cools can range from only 1°F to more than 5°F per 1000 ft. When the lapse rate (i.e., the rate of cooling) is small, the air is stabile. When the lapse rate is high, air is unstable.

When the air is unstable, a mass of air on the surface of the ground will be heated until its temperature is higher than that of the surrounding air. Hot air being less dense than cool air, the surrounding cooler air will squeeze the heated air upward. The warmer, thinner air parcel will then rise upward, gradually cooling as it gains altitude. This sort of rising air is called *convective* lift because it was generated by convective heating of the air by the ground. The individual pieces of warm lifting air are called *thermals.* The air temperature in a thermal cools as the rising air expands as its pressure gets lower with increased altitude. At some point, the air will cool to the dew point, and moisture in the thermal will condense out. The cloud thus formed is a nice, puffy, typical flat-bottom cumulus cloud such as you see on almost any summer afternoon over the mountains. Although triggered by a thermal, once they are established in unstable air, clouds of this type continue to build upon their own internal energy.

Thermals and the clouds they produce require less energy to form in the mountains. The sun heats the air at the same rate as, or even a slightly more rapid rate than, at sea level, but the cooler, thinner air heats more rapidly. Less heat is required to start the process. Less atmospheric pressure traps the hot air on the ground.

Here's a fun calculation: How high is the cloud base?

Well, you could ask other pilots, you could call Flight Service (see Chap. 2), you could observe where the cloud base lies in relation to mountains of known elevation, or you could try this:

Lapse rates.   The *dew point lapse rate* is about 1°F per 1000 ft during lifting. Translated to English, this means that the temperature and dew point get closer together at a rate of: $5^{1}/_{2}$°F per 1000 ft (the *dry adiabatic lapse rate*) − 1°F (the *dew point lapse rate*)= $4^{1}/_{2}$°F = the rate at which temperature reaches dew point per 1000 ft of altitude gain. Thus, if you know the surface temperature and you know the dew point temperature, you can calculate the cloud base. If the temperature at the surface is 50°F and the dew point is 32°F, the cloud base is 50 − 32 = 18°F. Divide 18°F (the temperature–dew point spread) by $4^{1}/_{2}$°F (the rate at which temperature approaches dew point), and you get 4. Since 4 × 1000 = 4000 ft, the cloud bases today are 4000 ft above the surface.

Convective lifting and cumulus clouds. Thermals are the bread and butter of lift employed by sailplanes. Glider pilots will discover they are in lift, partly just by the sudden sensation of going up, but more by reading their rate of climb instruments. They will circle in this thermal until they climb up to the cloud base. Then they will head out toward the next likely looking cumulus cloud, which is marking another thermal, and circle up under that cloud. How strong are these typical thermals under typical fair weather cumulus clouds? By the time cumulus clouds are just beginning to form, glider climb rates of 100 to 200 ft/min are typical. As the clouds grow in size, lift gets stronger. Virtually all glider pilots have routinely experienced sustained climbs in excess of 1000 ft/min under still-developing mid-afternoon cumulus clouds. As a rule of thumb, once the height of a cumulus cloud gets to be about 4 times its width, it is time to give it a wide berth in a small airplane.

The same sources of lift exploited by gliders can be used in the same way by pilots of small powered aircraft. If you don't have to, though, you will probably opt to avoid the light turbulence always associated with this lift. You may elect to fly above a layer of cumulus clouds because once the moisture condenses out, much of the energy is dissipated, so the air above these clouds is normally much smoother than the air below. Even if you can't climb high enough to fly above them, the air above the cloud base, between the clouds, is smoother than the air directly below the clouds. Slaloming between the clouds, within reason, can be a more comfortable (and possibly much more scenic) flight.

Some words of caution are in order here. First, flying above a cloud deck is very risky, never more so than in this situation. The clouds are still building, so cloud tops are still rising. Over the mountains, you could first become trapped above a solid cloud deck and then have the cloud tops keep rising until they were much higher than the maximum altitude your aircraft could attain. Now you are trapped above clouds over mountains. How and just where do you figure on descending through them? Second, developing cumulus clouds are on their way to overdeveloping. See the next paragraph.

About the time the height of a large cumulus cloud exceeds 5 times its diameter, it is likely seeing climb rates inside the cloud well in excess of 1000 ft/min, if not double that, but much more is to come. At that point, the moisture in the cloud may soon be raining out, which will produce downdrafts also in excess of 1000 ft/min. Just a tiny amount of bad luck is enough to allow you to encounter both in the same cloud, and the resulting turbulence is more than enough to damage a small aircraft. We have not yet reached the stage of a large thunderstorm, which adds still more turbulence, hail, icing, and lightning to the mix. And as you might remember, Dorothy, the largest cells of this type over the Great Plains spawn tornadoes. Listen on the radio the next time you fly on a day with thunderstorms—even large airliners deviate around them.

So a small cumulus cloud is not a threat in itself. If you encounter it late in the afternoon, it means you have a typical summer day. If you encounter it very early in the morning, it is often a sign of more severe weather on the way. Rapidly developing cumulus clouds during the morning flight to your destination could be a sign to turn around very quickly for the flight back home, or plan to wait for evening.

Both small fair weather cumulus clouds and the large thunderstorms they spawn are heat-driven. They build with the increasing heating of the ground from the morning sun. They dissipate in the early evening when the angle of the sun's rays gets lower. Midsummer afternoon thundershowers are the norm in mountains in the temperate regions; but clear, mild mornings and evenings are part of the same picture. When do *you* want to fly?

There is plenty of reason for pilots of small aircraft to stay clear of rapidly developing cumulus clouds.

Where cumulus clouds form

Standard texts on weather offer a few examples:

- *On a typical summer day* almost anywhere in the world, cumulus clouds will begin to form as a result of *ground convective heating.* The surface of the ground is heated unevenly. Lakes heat slowly, dry fields heat quickly, plowed fields and woodlands heat slowly, bare rocks and asphalt parking lots heat quickly. Air rises over the hot ground. Cumulus clouds form as the rising air cools. The sky is eventually covered with numerous little cumulus clouds, all about the same size with cloud bases all at the same altitude.

As these build, the afternoon can get very bumpy, but not dangerous for flying in a small plane. By evening when the sun's rays are more parallel with the ground, these clouds usually dissipate.

An advancing cold front. Typically cold fronts are quite easy to see. The cold air is causing condensation, marking the leading edge of the front. Aspen, Colorado, is somewhere about 10 mi farther on behind these clouds. Fletcher Anderson photo.

- *Advancing cold fronts* move across the surface of the ground in a wedge shape, as seen from the side. The air in front of the front is compressed and lifted. There is widespread cumulus buildup ahead of the front. This is called *prefrontal instability,* which sounds vaguely like the kind of mental disturbance which could be addressed with a lobotomy. Very, very large cells can develop over the Great Plains in these conditions, producing severe storms.

- *Air around low-pressure systems* is unstable and lifting, producing cumulus buildup and storms around a low. The larger of these are typhoons and hurricanes.

- *Over rising terrain:* What most texts neglect is the fact that terrain can also produce lifting and cumulus buildup. An air mass moving eastward across the desert will hit the Rocky

Mountains and be lifted. That mechanical lifting will also produce clouds. Thus, in the mountains, clouds form first over the high terrain. Major valleys will remain in the clear long after the nearby peaks have been completely obscured in clouds, rain, and lightning. This, by the way, is the reason mountains are generally wetter than the flatlands nearby. There is a desert west of the Rocky Mountains, and a rain shadow over the dry plains to the east, but the mountains themselves see considerable winter snow and summer afternoon rain.

The rugged terrain of the mountains automatically produces differential heating of the surface. One side of the mountain is getting direct sun; the other is in shade. Heated air will rise up the sunny side of the peak in unstable air and will continue rising to form a cumulus cloud. Even on a day with no weather predicted, the mountains themselves can generate local cumulus clouds, particularly in summer.

Other concerns with cumulus clouds

The strong lift and turbulence within cumulus clouds has already been alluded to. Presumably you don't intend to fly *into* a large thunderstorm. For obvious reasons, you would skirt it by at least a couple of miles.

Icing.   *Icing may* be found in cumulus clouds and will be examined in the chapter on IFR.

Lightning.   Lightning can be encountered not only under the storm, but near it as well. If we recall that cumulus clouds tend to form directly over high mountain peaks, it should be no surprise to learn that thunderstorms also form directly over these same peaks. Lightning tends to travel toward the nearest high ground, which is, once again, the highest peaks. Lightning does not automatically bring down a plane, although it can do so. Composite construction (just think fiberglass) gliders have literally exploded when the lightning traveled along the metal push rods which operate the ailerons and blew the skins off the wings. Large jets often have a tiny pinhole in the skin where the lightning hit, if they show anything at all. Smaller metal aircraft at best have all the circuit breakers tripped, and just as often all the instruments are fried. Local pilot Peter Lert described his most recent (of many—he has been flying in all weather for over 30 years professionally) lightning strike in a turbo Commander as a sound like the discharge of a 410 shotgun in the cockpit. He equated the effects as more like the sound of a very large, old-fashioned cash register ringing up the bill. This is a particularly apt analogy, as the bill to replace all the radios was very significant indeed.

A snow microburst. The vertical white plume of snow marks a severe, but very narrow downdraft. Fletcher Anderson photo.

Microbursts.   Once the cloud overdevelops, some of the condensed out rain and hail in the cloud begins falling. Downdrafts of unbelievable velocity can be found in mature-stage thunderstorms. Two or three thousand feet per minute has been measured often enough to be considered routine. Twice that has been estimated or seen with Doppler radar. I have personally encountered 3000 ft/min down a mile out on final to Telluride.

Virga and mammata.   Microbursts are often visible as *virga* (see below). *Mammata* are mammary like bulges on the bottom of a developing cumulus cloud. They mark downdrafts within the cloud and are a precursor of both precipitation and microbursts. Microburst events are a normal part of thunderstorm development, not something unusual at all. At least in the arid west, you can observe these under clouds and simply fly around them. In a humid climate, thunderstorms and their attendant microbursts can be embedded in the overall haze, and you can't see them.

Encounters with microbursts when on final approach have brought down large jet airliners. Assuming the microburst is somewhere just ahead, the pilot feels a strong, abrupt gust on the nose and reduces power to stop the sudden very strong lift. Next the pilot flies into the very strong downdraft of the microburst itself and adds full power to stop the descent. This is bad enough, but more is still to come as the planes flies out the other side of the downdraft. Now the plane is too low and very slow in a climb attitude. Leaving the downdraft, the pilot experiences a very severe gust on the tail, which he perceives as still stronger sink and a very abrupt loss of airspeed. The plane can either stall or simply settle into the ground. Investigations into microbursts began in earnest following the August 2, 1985, crash of Delta flight 191, a Lockheed L1011 jumbo jet in Texas. Doppler radar was then in its infancy and was vital to the investigation. Before Doppler radar, there was no reliable way to acquire the data.

How do you avoid microbursts? You avoid flying right under a cloud with mammata, and you don't fly through virga. The visual clues are there to read. Mostly, though, you simply give any large thunderstorm a few miles clearance.

The strong downdraft of a microburst produces two other phenomena: virga and gust fronts.

Virga is sometimes defined as rain that evaporates before it hits the ground. Virga is also the result of cold air coming down out of the cloud, causing condensation as it hits warmer air below. Virga appears as vertical or near vertical streaks coming down out the bottom of mature cumulus clouds. When you see virga, you are *seeing* the downdraft or microburst. Don't fly through it!

Gust fronts. Gust fronts occur when a downdraft or microburst hits the ground. A good analogy is the stream of water from a garden hose directed at the ground. The water hits the ground and spreads out in all directions. Cold air descending from the cloud is denser than the surrounding warm air and also spreads out over the surface. Seen in cross section, it resembles a miniature cold front. From the vantage point of an aircraft, you can often observe the rolling dust cloud as one of these gust fronts moves across the ground. A thousand feet or more above it, what you feel is usually the rising air wedged up by the gust front moving along the surface. Being hit from behind by the gust front when taking off or landing would mean very demanding turbulence associated with wind shear, which could include a sudden drop in airspeed large enough to cause a stall.

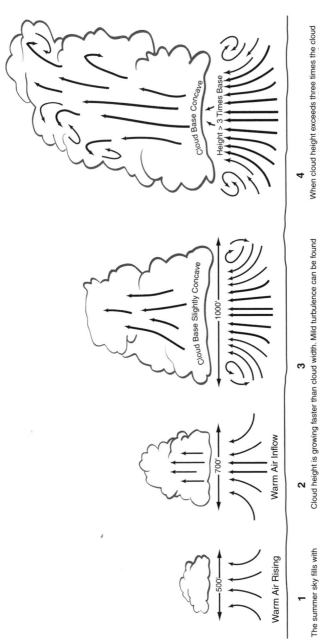

1

Warm Air Rising

The summer sky fills with numerous small clouds. These are of no concern to pilots, but of great interest to glider pilots who will circle beneath them in lift.

2

Warm Air Inflow

Cloud height is growing faster than cloud width. Mild turbulence can be found close to and under these clouds, smoother air above. Clouds #1 and #2 typically mark the tops of thermals rising from the ground.

3

Cloud Base Slightly Concave

4

Cloud Base Concave

Height > 3 Times Base

When cloud height exceeds three times the cloud width, small aircraft will do well to stay out from underneath because of strong lift and moderate turbulence.

The concave base of clouds #3 and #4 occurs about the time the cloud begins to develop its own internal energy and is no longer just fed by thermals.

Cumulus cloud development.

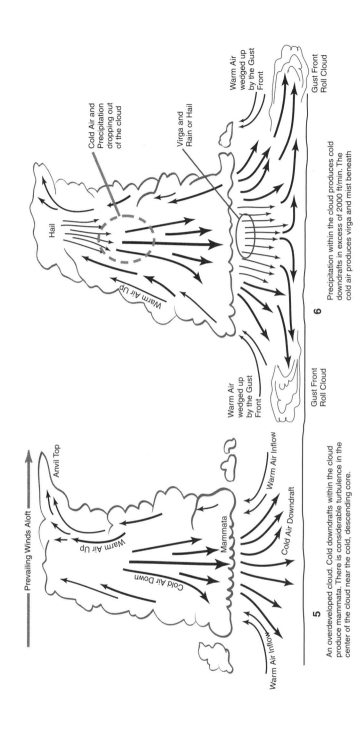

Prevailing Winds Aloft →

Anvil Top

Warm Air Up

Mammata

Cold Air Down

Warm Air Inflow

Cold Air Downdraft

Warm Air Inflow

5 An overdeveloped cloud. Cold downdrafts within the cloud produce mammata. There is considerable turbulence in the center of the cloud near the cold, descending core.

While cloud #4 can have lift in excess of 1000 ft/min, cloud #6 can have downdrafts in excess of 2000 ft/min, i.e. 20 knots of wind blowing straight down.

Cold Air and Precipitation dropping out of the cloud

Virga and Rain or Hail

Hail

Warm Air Up

Warm Air wedged up by the Gust Front

Gust Front Roll Cloud

Warm Air wedged up by the Gust Front

Gust Front Roll Cloud

6 Precipitation within the cloud produces cold downdrafts in excess of 2000 ft/min. The cold air produces virga and mist beneath the cloud. This downdraft is a microburst, which spreads out to form a gust front.

83

In flat country, gust fronts soon dissipate. In the mountains, gust fronts can be channeled by valley walls and run for very considerable distances. In the days when I taught paragliding in Aspen, Colorado, we would observe virga under clouds over 40 mi away near Mount Sopris. An hour later, we would feel the very sudden passage of a gust front with winds over 20 knots. Under optimal conditions, we could actually watch the wall of dust and debris roll up the valley toward us. It was important that we figure this out—a paraglider's maximum speed is less than 30 knots. The result of an encounter with one of these fronts for a paraglider could mean being blown back over the summit of the mountain into potentially unsurvivable turbulence.

Over flat country, putting at least 5 mi between you and even a small thunderstorm would not be excessive. Over mountainous terrain, considerably greater distance than that would do no harm at all.

The urge to hurry in and land right before the storm hits runs the risk of arriving just as the gust front does. Most of us have made this mistake at least once. Thunderstorms are normally very localized phenomena, and gust fronts and microbursts are typically extremely short-duration events. Flying out to the wide part of the valley and circling for as little as 10 min to let the storm go by are a safer bet than arriving seconds before it does.

Cumulus cloud types

Cumulus clouds can be classified into several basic types:

Fair-weather cumulus.   These clouds are the puffy little afternoon clouds seen almost every summer day in the mountains. They are often described as looking like little puffs of wool. All the clouds are about the same size, and cloud bases are all about the same altitude. They are the result of localized lifting and are typically 2000 or 3000 ft or more above the surface. They are often a precursor of afternoon thundershowers. Light turbulence is generally associated with these clouds. Once they form, they generally continue building until near sunset. Then the sun's rays are hitting the earth at a flatter angle; there is less convective heating, so the source of the convective lift that formed the clouds is gone.

Fair weather cumulus clouds, while they are still small, are of little concern to a pilot. When you observe these clouds, the rate at which they are building should help you judge whether you have time to reach your destination. You might also consider that a return flight

Fair weather cumulus clouds. Small clouds like these build up almost every summer afternoon. Although the air is somewhat bumpy, at this stage there is little to worry about. Fletcher Anderson photo.

home the same day might still be very reasonable in the late afternoon or early evening once these clouds have dissipated.

Cloud streets.   These are long lines of small cumulus clouds. There is a band of lifting air under each line of clouds and a corresponding band of sink in the clear air in between. Gliders head out in a straight line under these streets, slowing down in the lift under each cloud and speeding up in between. While you can do the same in a small plane, the ride is probably more comfortable if you just push on through the clear air gentle sink between the clouds.

Cloud streets often align themselves over a ridge or range of mountains. A true cloud street is a long line of cumulus clouds all drifting downwind in a row, all having the same point of origin at the earth's surface.

Altocumulus.   These clouds are like fair weather cumulus clouds, but higher (*alto* means high) and much more densely packed. Cloud bases are typically 10,000 ft above the ground. There can be some turbulence associated with these clouds. What you are probably seeing is the overall lifting of the air in advance of a warm front. You should be expecting rain or snow anywhere from a few hours later to the next day. You can frequently correlate these clouds with the predicted approach of a front.

The presence of slowly developing altocumulus clouds does not mean that flying is a bad idea right now; it means bad flying weather could be arriving in a few hours.

Cirrocumulus.   These clouds are higher still, with bases of 20,000 ft or higher. *Cirrus* in cloud speak means very high. These can be either the lifting of an incoming front or the boundary layer between two air masses. These clouds can have a wavy appearance and are sometimes called *mackerel sky*. They often suggest weather coming in over the next couple of days, but do not necessarily portend anything of concern to a pilot over the course of the next couple of hours. In fact, they probably mean better flying weather over the next couple of hours because the thin overcast layer will shut off some of the sun's rays and slow convective heating and the resulting turbulence.

Cumulonimbus.   These clouds are rain clouds. *Nimbus* means rain. As a cumulus cloud grows, it becomes a towering, billowing dark mass of cloud which eventually begins producing rain, often very heavy rain. This type of cloud can reach all the way from the surface up to over 30,000 ft. Sometimes violent turbulence is associated with these clouds. You aren't planning to fly through *or under, or close by* around these clouds in a small airplane, right? Flight Service is going to advise you not to. They will issue a *convective sigmet* to tell you so. Heavy rain pulls down air as it falls, so flying through heavy rain means strong sink as well as poor visibility. At mountain altitudes, your rate of climb is already poor, so you will elect not to fly through rain which would be no trouble at sea level.

The usual sources of these clouds are regular summer afternoon convective buildup or the wedging up of unstable air by a rapidly advancing cold front. In either case, you are looking at a weather event that will likely have passed by the next day.

Note that mountains themselves produce this situation and cloud type very regularly in the summer months. Air masses hitting the mountains are forced upward by rising terrain. If the air has any instability, the result is rapid development of thunderstorms. Examining the rainfall map for any mountainous region will show you that the upwind side of mountain ranges gets substantially more precipitation than the downwind side or the flat country on either side. You can observe the results as you fly over. West of the Rockies lies the desert. The Rocky Mountains

Late-afternoon thunderstorms. A small, or even most large, aircraft would do well to avoid these thunderstorms. Notice that while the mountains are obscured, the surrounding areas are not. Fortunately if you have to get somewhere, these storms tend to dissipate as the angle of the sun gets low in late afternoon or early evening. Shelby Evans photo.

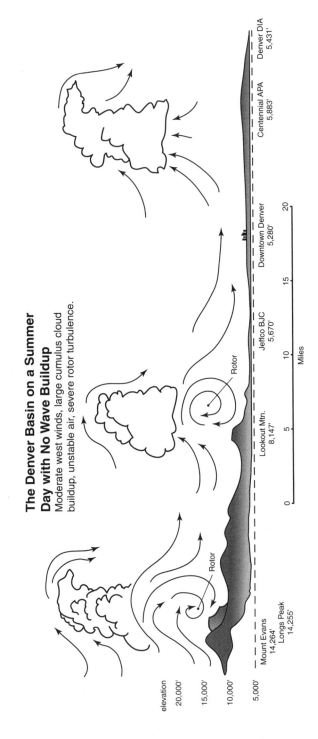

The Denver Basin on a Summer Day with No Wave Buildup

Moderate west winds, large cumulus cloud buildup, unstable air, severe rotor turbulence.

THE DENVER BASIN LOOKING NORTH

themselves are heavily wooded and green up to timberline. The plains east of the Rockies are yellow short-grass prairie dry enough that 100 years ago they were incorrectly referred to as the Great American Desert. The rain forests of the Cascade Mountains in the northwest and the Coast Range in British Columbia are just upwind of very dry eastern Oregon, eastern Washington, and the dry prairies of central Canada.

Stratocumulus. These clouds are in layers (*stratus* means layer). These are seen as the breakup of a stratus layer due to decreasing air stability. What was a fairly solid layer of overcast is turning into a broken layer of cumulus. This *could* mean that you are witnessing the breakup of clouds of residual moisture after a passing storm when the sun burns them off. You are seeing rapidly improving weather, and it is the perfect moment to fly. *It could also mean that a new system is on the way in* and that you are seeing the result of prefrontal lifting. It is an almost perfect time to fly right now, but things will deteriorate soon and you have a very small window of opportunity. Without additional information, it is very difficult to evaluate these conditions. This might be an excellent time to call Flight Service because they will know if a large system is moving into the area. The afternoon trip home might see very early thunderstorm buildup.

Cumulus fractus. These very ragged clouds are being literally blown apart by very high winds. This is an obvious indicator of severe turbulence as well. The pronounced slant of the cloud tops indicates the wind is blowing in that direction. Fletcher Anderson photo.

Fractocumulus.   These clouds are cumulus clouds being blown apart with very feathery edges. *Fractus* means broken. They are being blown apart by very high winds and turbulence. In all probability, if you are not experiencing turbulence yet, you soon will be, and at all altitudes as the strong winds reach lower altitudes.

The photographs of fractocumulus clouds show another phenomenon—*cloud lean*. The clouds build vertically. The tops rise up into stronger winds, and the cloud acquires a pronounced lean in the direction in which the wind is blowing. Knowing wind direction can help you avoid the downwind side of ridges and mountain ranges.

Typical cloud heights

Cumulus, and indeed all, cloud elevations vary greatly with temperature and therefore with the seasons. Warmer air holds more moisture than colder air, and summer air is warmer than winter air. Summer cumulus clouds generally have cloud *bases* well above most mountain summits. Winter cumulus clouds in the Rocky Mountains often have cloud *tops* close to or even occasionally below the highest mountain summits. It is unlikely that you could get high enough in a small plane to fly above summer cloud tops in the high mountains, but quite likely that you could cross most major mountain passes below cloud base (at least until it started raining later in the day). In winter, cloud buildup can and usually does obscure the mountains and often closes off the passes you hope to fly through. On the other hand, many small aircraft are quite capable of flying at altitudes from 14,000 to 18,000 ft in VFR conditions above the winter cloud tops. A word of caution, though: You don't want to try this unless you are very comfortable in your ability to descend in IFR through the cloud layer which could quickly rise up to engulf you.

Cumuluslike mountain cloud types

These are clouds which at a glance look very much like normal cumulus clouds. In fact, they could well be found together with regular cumulus clouds and can be very difficult to distinguish from them. They are discernible in part because typical cumulus clouds form and then drift downwind, while clouds in this category are essentially stationary.

Banner clouds.   These form on the lee side of mountains in strong winds. These can be cumuluslike in shape, but are the result of cooling due to lower air pressure in the lee of the mountain. Seen from

Banner Cloud in the Lee of the Mountains

The winds blowing right to left in this illustration have created a vertical eddy north of the mountains and the banner cloud is a stationary cumulus cloud in this eddy.

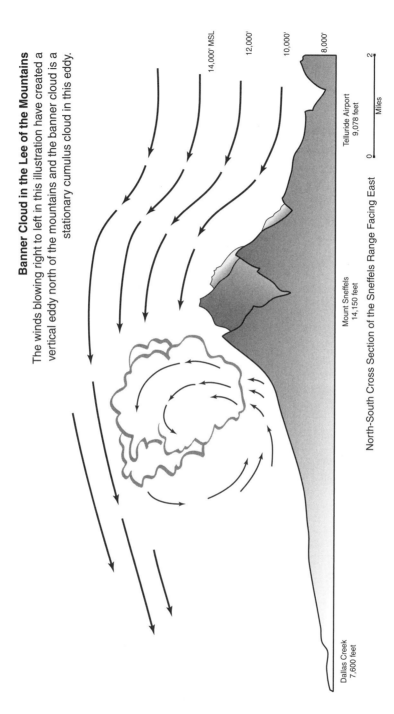

14,000' MSL

12,000'

10,000'

8,000'

Telluride Airport
9,078 feet

Mount Sneffels
14,150 feet

Dallas Creek
7,600 feet

0   Miles   2

North-South Cross Section of the Sneffels Range Facing East

Blowing snow. Looking south along the length of the 14,000-ft-high Collegiate Mountains in central Colorado. Surface winds reported by the newly installed AWOS on Monarch Pass in the middle distance were gusting over 70 knots. At our altitude of 17,500 ft we have an 85-knot crosswind for our direct route from Telluride to Denver. By selecting a route upwind from these and other major mountain ranges, we experienced only light turbulence. Lower down or on the lee side of one of these ridges, turbulence would be fatal in a small plane. Fletcher Anderson photo.

the side, they resemble an extension of the upwind side of the mountain. In the winter, a banner cloud is often a mixture of cloud and fresh snow being blown off the upwind side of the peak. An illustration is better here than a description.

The banner cloud should alert you to the presence of strong wind. The lee side of a ridge is no place to be in those conditions. The upwind side could be very smooth.

A roll cloud or rotor cloud. This forms, as a banner cloud does, in the lee of a ridge. It is a horizontally rolling cylinder of air in which you can actually perceive a rolling or tumbling motion in the cloud. You are actually seeing the turbulence. Rotor produces a combination of sink and turbulence strong enough to be lethal in a very small, light plane. Rotor is less likely on the lee side of a very gradual ridge, but almost a certainty on the lee side of a sharp, steep ridge. (See the illustrations on p. 96.)

Rotors and rotor clouds can also form underneath mountain wave, as shown in Brian Lewis's photograph. The first photograph shows a whole row of horizontally rolling cylinders of air, each directly beneath a wave crest and each marking an area of turbulence.

Banner cloud. A 20-kn wind is blowing snow off Mount Ballard, Colorado, even though on this winter day, the wind is calm on the valley floor below a temperature inversion. Fletcher Anderson photo.

Summer and winter views of a banner cloud forming on the lee side of Dolores Peak, Colorado, as seen from the runway at the Telluride Airport. These are indicators of a high wind coming from behind the peak. Ron Kanter photos.

The Telluride Airport. Notice the steep cliffs along the south side of the runway. The previous two illustrations both indicate strong south winds, which would blow up the cliff and produce rotor turbulence as well as a crosswind on the runway. Ron Kanter photo.

Billow clouds. Billow and roll clouds are horizontally rotating cylinders of air essentially similar to rotor clouds, but not as directly associated with the lee side of a ridge. They can form in the shear layer between two air masses or under a lenticular cloud.

Valley wall clouds. These are conventional cumulus clouds glued up against the upwind side of a ridge. They are generally found when the wind is light and there is considerable moisture in the air, for example, when there is considerable residual moisture after a storm the evening before. They form when light wind deflects the air upward as it encounters the ridge. Very light lift and minimal or no turbulence are associated with these clouds. They might be growing, but if seen very early in the morning, they may be burning off as drier weather comes in. Their principal difficulty for a pilot is that they obscure the mountains. Wider valleys may stay open for some time, but getting high enough to cross mountain ranges or finding passes with windows to duck through may be problematic. Here is another great

Roll clouds. What look like rows of cumulus clouds are in fact stationary roll clouds. These are horizontal rolling cylinders of air. They mark areas of turbulence. Arranged like this in rows, they probably lie underneath mountain wave. Brian Lewis photo.

Rotor clouds. What look a lot like cumulus clouds are in fact stationary, not blowing off downwind. The cloud is a rolling horizontal cylinder of air, marking an area of local turbulence. Notice also how turbulence beneath the rotor is tumbling the smoke of a fire in the left foreground. Brian Lewis photo.

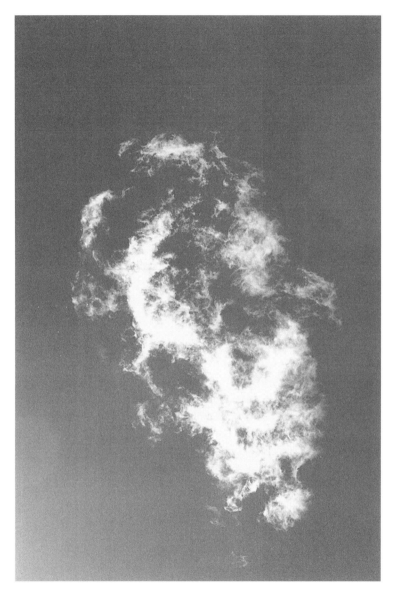

Billow cloud. Seen as a still photograph, this looks like an early stage of development of a cumulus cloud. In fact, it is displaying a rapidly rolling motion, easily discernible either in flight or from the ground, and it is a marker of localized turbulence. Several of these forming and disappearing all at the same altitude mark a shear layer. Fletcher Anderson photo.

Valley Wall Clouds

Early morning upslope winds are lifting air up the valley wall until it cools to form a stationary cumulus cloud. These are typically associated with residual moisture after a storm the day before.

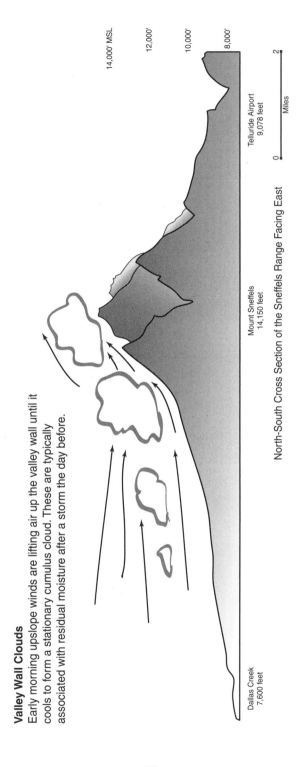

North-South Cross Section of the Sneffels Range Facing East

Dallas Creek
7,600 feet

Mount Sneffels
14,150 feet

Telluride Airport
9,078 feet

14,000' MSL

12,000'

10,000'

8,000'

0   2
Miles

98

time to check with Flight Service before departing, because they will be aware of overall drying or deteriorating trends.

Stratiform clouds

Stratus means layer. A stratus cloud is therefore a flat layer of clouds. Low-level stratiform clouds generally mean stable air and no or very light winds. For a pilot, the concern with this type of cloud is primarily visibility. Icing can be encountered in stratiform clouds and is dealt with in the chapter on IFR.

Stratus clouds. Stratus clouds are a solid layer of overcast clouds, usually lower than 6000 ft above ground level (agl) and not necessarily very thick. They tend to cover large areas and may have light, drizzling rain. Once settled in, this type of weather pattern is not likely to rapidly change, even in the mountains. This is the type of rain or snow associated with a warm front.

Here is an interesting factoid from the southwestern desert: The Navajo refer to this as "female" rain and the rain associated with big cumulus buildup as "male" rain.

Valley wall clouds. Clouds like these are typically formed of residual moisture from a storm the day before. They form in rising air on the upwind side of the mountains. Strong winds would blow them away, so provided you don't fly through them and hit the mountains behind, they are typically a sign of very light wind and no turbulence. Shelby Evans photo.

Valley wall cloud. The cloud on the left side of the frame is a valley wall cloud. John Kounis photo.

Nimbostratus.   This is the same thing, only with rain. *Nimbus* means rain.

Both cloud types above can be associated with an incoming warm front overrunning cooler air closer to the ground. Thus you could have rain falling from these clouds into air colder than the freezing point below. These rain drops become supercooled and can quickly freeze as soon as they hit something solid to adhere to, such as your wing. Here is another situation in which a call to Flight Service inquiring about icing might be in order. In general, freezing-temperature air below a light drizzle causes very rapid buildup of smooth airframe ice. These cloud types often occur in multiple layers, and a descent through them can produce mixed icing.

In warmer conditions safely above the freezing point, poor visibility is the only concern, and sudden encounters with much worse visibility come more from additional clouds forming in more layers than from increased rain. Over flat country, these are conditions in which you might push on in bad weather. In the mountains, you would not, because terrain is hard to spot early enough and because localized lifting could produce locally worse visibility quickly, even with an overall stabile system.

Altostratus.   These clouds are higher, typically 10,000 agl. For the pilot these pose little hazard. The air is typically stable, there is no rain, and there is little or no turbulence. There may well be wet weather on the way in association with a warm front, but not necessarily in the next couple of hours before completion of the flight.

Cirrostratus.   These clouds are very high, thin layers of cloud at altitudes of 20,000 ft and up, sometimes up to 30,000 or 40,000 ft msl. These clouds are often a predictor of a cold front arriving the next day, but are of little immediate concern to a pilot.

Cirrus.   These clouds are very thin, wispy clouds at very high altitudes (well above 25,000 ft). They are much the same as cirrostratus clouds, except that there is less moisture aloft to form clouds. They, too, could be a predictor of a cold front a day or two away. The moisture itself may have been lifted to high altitudes by large thunderstorms associated with the front and then blown far out in advance of it by higher winds at very high altitude.

Mountain weather expert David Whiteman has observed that thin, wispy high cirrus clouds over the mountains of Colorado often began life as the detached anvil tops blown off of large thunderstorms over the Mogollon Rim north of Phoenix, Arizona. Margaret Lamb has named these "Little Orphan anvils." In Colorado, they can be indicators of a southwesterly flow aloft, which might be a precursor of moisture coming up from the Sea of Cortez—or they might mean nothing at all.

Temperature inversion layers

Inversion layer clouds are very common in the mountains, and they could also be classified as a form of valley fog. Air cools at night. As it cools, it becomes denser and therefore heavier. The cold air flows downhill into the valley bottoms. The normal situation is that air gets cooler as you get higher. The rate of cooling is called the *lapse rate*. With a temperature inversion, the actual lapse rate is much slower than the normal lapse rate. There can even be cold air below warm air. When this situation occurs, if there is moisture in the air, it will be trapped under the inversion layer and form a stratiform cloud very low to the ground. Particularly in winter, and particularly after cold, clear nights when heat radiates away readily, all mountain valleys experience temperature inversions. So do major basins just outside the mountains such as Denver and Salt Lake City. In these cases,

The jetstream. On the day that this photograph was taken, the jetstream with winds of 130 knots was reported directly overhead at 40,000 ft. This cloud probably marks the boundary of those winds and calmer air below. Because of local temperature inversions, the air below is completely calm. Fletcher Anderson photo.

Temperature inversion. The pronounced haze in this picture is very mild air pollution trapped under the temperature inversion layer in the Telluride Valley. The air is extremely stabile, and pilots Bob Wallick and Joline Esposito are enjoying a very smooth, if rather cold, flight. Ron Kanter photo.

the cloud is not just moisture, but everything else that got into the atmosphere as well. In Denver or Salt Lake City, it will be a cloud of automobile emissions. In the 1960s, most older homes in mountain communities in Colorado still had coal-burning furnaces left over from the mining era. Camp Hale, Colorado, at 10,000 ft where the mountain troops trained in World War II had coal heat plus coal-burning railroad locomotives. At times more than one-half of the troops were incapacitated with respiratory problems.

These cloud layers are very thin. Often cloud tops are only 100 or 200 ft above the valley floor, rarely over 1000 ft. Flying over the mountains in winter on a crystal-clear day, you can look down early in the morning and see Eagle and Gunnison, Colorado, Jackson Hole, Wyoming, or South Tahoe, California, or other similar airports in relatively closed

Temperature inversion. Stagnant air under the temperature inversion is trapping all the pollution in the Denver area, forming a brown cloud typical of that city in winter. The top of the brown cloud is only 1000 ft above the surface. Wind at the altitude from which this photograph was taken is above 20 knots. Surface winds are calm.
Fletcher Anderson photo.

valleys with narrow outlets completely socked in with solid IFR conditions, but with cloud tops only a few hundred feet above the ground.

Flying into Denver or Salt Lake City in winter, you will see the entire basin filled with a decidedly unhealthy-looking brownish yellow cloud, but the tops of the highest buildings downtown rise up through it into clean air. If you are president of the company, go for the top-floor office! Residents pray for a good high wind or snow shower to clear out the air.

To simply shut your eyes and bust up through these layers is ill advised because of the proximity of the surrounding mountains. For appropriately qualified pilots, an IFR departure through the clouds to VFR on top may make excellent sense. Departing Jackson Hole, Wyoming, or South Lake Tahoe, California, you could depart in dense, gloomy fog and break out into bright sunshine in as little as 30 s into the flight. Just waiting works, too; the sun will eventually burn off this kind of cloud.

Temperature Inversion in Mountain Valleys
Cold air settles into the valley at night.

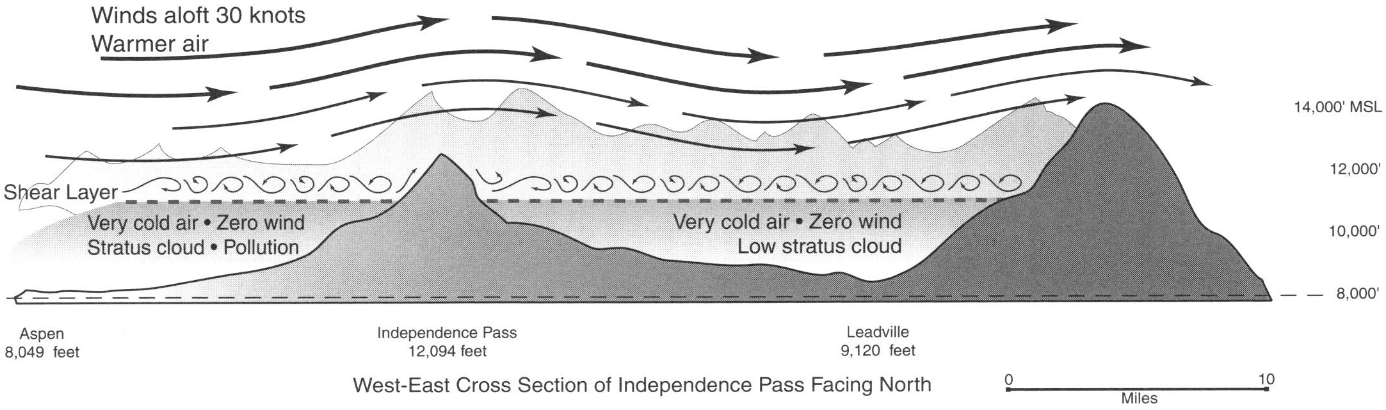

Winds aloft 30 knots
Warmer air

Shear Layer

Very cold air • Zero wind
Stratus cloud • Pollution

Very cold air • Zero wind
Low stratus cloud

14,000' MSL

12,000'

10,000'

8,000'

Aspen
8,049 feet

Independence Pass
12,094 feet

Leadville
9,120 feet

West-East Cross Section of Independence Pass Facing North

0   10
Miles

105

Residual moisture after a storm. The morning sun will quickly burn off this thin layer of low clouds, which mark a typical short-lived summer morning temperature inversion. John Kounis photo.

Morning fog. Willits Airport, north of San Francisco, California, is experiencing solid instrument conditions, except for this convenient little blue hole, even though the morning fog layer is only 150 ft thick. The sun will soon burn this off. John Kounis photo.

Lenticular clouds

Lenticular clouds are lens-shaped with relatively hard, smooth edges. (Perhaps more accurately, they are lentil-shaped—*lenticularis* is the Latin word for lentil.) They remain stationary in the sky, which may seem paradoxical because they are indicators of very high winds. A lenticular cloud is formed by the cooling of lifted air. The cloud is constantly forming at the upwind edge and disappearing at the downwind edge as the air descends and warms. A stack of lenticular clouds resembles a fleet of giant flying saucers. Lenticular clouds may be seen in three distinct situations. In all three cases, the clouds themselves are telling you the same thing: High winds are being lifted up and then dropping back down. Since high winds are producers of strong turbulence, pilots should see lenticular clouds as a cause for concern.

Cloud caps (or Foehn walls in Europe)

These are lenticular clouds forming directly above the summits of mountain peaks. Very strong winds are deflected up the upwind side

Lenticular clouds. These hard-edged stationary clouds are markers of high winds. They also frequently mark the crests of mountain wave.
Brian Lewis photos.

of the mountain and then sink down the back side. A cloud cap normally forms about 1000 to 3000 ft above the top of a peak, but much, much higher is not unusual, nor is lower. Cloud caps 40,000 ft above the Rockies are a regular occurrence. The exact height varies greatly with temperature and moisture. You will frequently

Lenticular cloud stack. Several lenticular clouds in a stack are a sure marker of wave lift. Note also that regular cumulus clouds are forming in the same lifting air. Brian Lewis photo.

Lenticular cloud cap. The lenticular cap over a mountain is an indicator of strong winds. Brian Lewis photo.

Cumulus fractus. Cumulus clouds are trying to form, but strong winds are shredding them into a fine mist. This is a clear indicator of very strong wind. Notice a lenticular cloud above, also a high-wind indicator. Fletcher Anderson photo.

see stacks of several lenticular clouds. The Foehn wall is a low lenticular cap over the whole of a mountain range, as shown in the Brian Lewis photograph on p. 112. Foehn winds in Europe and Chinook winds in North America are accelerated downslope winds of considerable strength found on the lee side of the mountains. Surface winds in excess of 60 kn can be encountered in certain localities near the mountain front, such as Boulder, Colorado. Fortunately for pilots, the National Weather Service is able to predict Chinook winds very accurately. The presence of the Foehn wall cloud is a sign to call them.

Pileus clouds

These are the same thing forming as a cloud cap over the top of a towering cumulus cloud. The cumulus cloud is acting as a mountain pushing up into the oncoming winds aloft.

Mountain wave

Lenticular wave clouds mark the crests of *mountain waves* forming downwind from a mountain or mountain range. Typical mountain

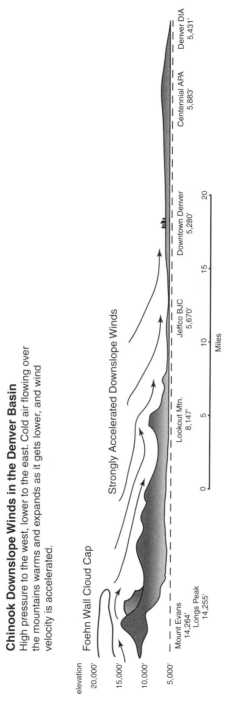

Chinook Downslope Winds in the Denver Basin

High pressure to the west, lower to the east. Cold air flowing over the mountains warms and expands as it gets lower, and wind velocity is accelerated.

Foehn Wall Cloud Cap

Strongly Accelerated Downslope Winds

elevation
20,000'
15,000'
10,000'
5,000'

Mount Evans
14,264'
Longs Peak
14,255'

Lookout Mtn.
8,147'

Jeffco BJC
5,670'

Downtown Denver
5,280'

Centennial APA
5,883'

Denver DIA
5,431'

0   5   10   15   20
Miles

THE DENVER BASIN LOOKING NORTH

111

Foehn wall. Brian Lewis's photograph of the Foehn wall cloud cap over the Colorado Front Range looks a lot like what it really is—a waterfall of very cold air spilling over Longs Peak down into the Denver basin. As it heats up and expands, the wind velocity is increased and a very strong Chinook wind is the result. Brian took this photograph out the window of his glider from about 18,000 ft on his day off from his regular job as an airline pilot. Brian Lewis photo.

wave lenticular clouds are found at altitudes from 20,000 to 40,000 ft msl, but lower is by no means unusual. The lowest cloud can often be at the same altitude as the mountains themselves. *Mountain wave* is a phenomenon of considerable consequence to pilots of large and small aircraft alike. Mountain wave is the bogeyman of mountain weather. A lot of wonderfully evocative, romantic writing about mountain waves breaking against the barrier of the peaks has been published. This is absolute garbage. Mountain waves are not in any way like ocean waves that break on the beach. They are exactly like river waves, which are *stationary,* remaining in one spot as the water rushes through them. Brian Lewis's photograph of smoke trapped in a mountain wave on p. 113 shows the true form. If you already know what a mountain wave is, then lenticular clouds can be a useful sign helping you to detect it.

Whenever lenticular clouds appear, there are high winds aloft. Conventional advice simply warns you not to fly at all whenever you see

Smoke in a mountain wave. If not the most artistic, this is certainly the most explanatory photograph of mountain wave ever taken. Brian Lewis shot this off his deck one afternoon as smoke from a fire in the foothills west of Denver was carried into mountain wave. Although the wind aloft is blowing about 30 knots, the waves themselves are stationary, as are river waves. They do not travel along as ocean waves do. Compare this photograph with the illustration on page 114. Brian Lewis photo.

them. As with all conventional advice, there is a core of truth to this assertion. Directly beneath, particularly directly beneath the down-wind edge of a lenticular cloud, you can encounter severe turbulence which may or may not be marked by a rotor cloud. The winds producing the lenticular cloud are definitely strong enough to produce turbulence on the downwind side of a peak with a cloud cap. A towering thunderstorm powerful enough to deflect the winds above itself and form a pileus cloud is in and of itself powerful enough to be producing severe turbulence.

Yet consider this: Back in 1961 Paul Bickle broke new ground and the world record when he climbed to an altitude of 46,200 ft in wave lift and lenticular clouds over the Owens Valley of California, east of the Sierras in what today would be considered to be a very basic metal Schweitzer 1-23 glider. (Modern composite gliders have been to over 50,000 ft.) In October 2000, German pilots flew an over-loaded Stemme motor glider over 1700 mi on one flight at very high

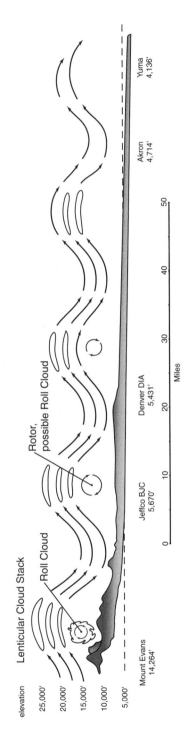

Mountain Wave East of the Colorado Front Range

Wave develops with winds aloft over 30 knots and a strong wind gradient with altitude, high pressure, stable air.

On this hypothetical day, the wave above 18,000 feet would have strong up- and downdrafts, but no turbulence because airflow is laminar.

Below 17,000 feet, severe turbulence could be encountered in rotor.

elevation Lenticular Cloud Stack

Roll Cloud

Rotor,
possible Roll Cloud

25,000'
20,000'
15,000'
10,000'
5,000'

Mount Evans
14,264'

Jeffco BJC
5,670'

Denver DIA
5,431'

Akron
4,714'

Yuma
4,136'

0   10   20   30   40   50

Miles

THE DENVER BASIN and the FRONT RANGE LOOKING NORTH

114

Lenticular wave clouds. These mark the wave crests, as in the illustration on page 114. Brian Lewis photo.

altitudes in mountain wave marked by lenticular clouds east of the Andes Mountains with the motor off. The older glider and the modern Stemme (because it was carrying much more than its designed load) would have been at risk in severe turbulence. How did they do it? A closer look at the phenomenon of mountain wave is in order.

Envision a rock in a fast-flowing stream. Water wells up against the upstream side of the rock, flows over it, and sinks down on the back side. You have the crest of a wave upstream from and directly over the rock and a wave trough just downstream from the rock. Just downstream from that, the water is rebounding to form another crest, then farther downstream falling back to form another trough. You get a series of standing waves downstream from the rock. Water is flowing through the waves, but the waves themselves are stationary. Precisely the same thing occurs with wind blowing over a mountain. A series of stationary waves sets up downwind. The wind is blowing quite hard, but the waves are stationary. Lenticular clouds form where the air is lifted at the wave crests. The lenticular clouds themselves are stationary even though the wind is blowing through them. Moisture condenses at the upwind leading edge of the cloud as it is lifted and cools, and moisture goes back into suspension as it descends and warms at the downwind edge of the cloud. Lenticular clouds may or may not be present with mountain wave, depending on the available moisture in the air.

Generally speaking, the upwind edge of a lenticular cloud has strong wind, strong lift, but very little turbulence. Under the downwind edge of a lenticular you will encounter strong sinking air and will likely—but not always—encounter turbulence. The air flowing through the wave is in smooth, laminar flow, as the smooth-edged clouds suggest.

Underneath a lenticular, particularly under the lowest cloud in a stack, you may encounter considerable turbulence. Given enough moisture in the air, there may be cumulus clouds associated with the lifting part of the wave. Underneath the lowest lenticular cloud in the stack, you could see what looks like a little cumulus cloud, but is actually a rotor or rolling cloud marking turbulence.

The concern with severe turbulence in wave conditions is not up high in the waves themselves. It is in the lee side of the ridges that form the waves and the possible cutoff rotors beneath them, as well as near the ground in the very strong surface winds often found in association with the even stronger winds aloft.

Conditions favoring wave formation

High winds alone do not necessarily create mountain waves. If the air is unstable, for example, when a front is coming in, then wind lifted as it hits a mountain wants to just keep on going up. There will be considerable disorganized turbulence and sink on the downwind side of the mountains, but no wave. Ideal wave formation conditions require wind direction close to perpendicular to the mountain range or no more than 30° from perpendicular. Wind speed must be at least 15 kn at mountaintop height, if not much stronger, and the wind velocity should increase greatly with altitude. The air mass should be stabile, as with high pressure. Ideal conditions would also include a temperature inversion at or near ridge top height, or at least below 15,000 ft.

Wave dimensions

How big a phenomenon are we talking about? Wave dimensions can vary greatly.

Many pilots in Colorado have encountered smaller wave trains within the major mountain valleys such as waves north of the San Sophias over the Montrose Valley, waves east of the San Juans over the San Luis Valley, or waves over South Park east of the Collegiate Range. These waves may have a wavelength (distance from crest to crest) of as little as 2 mi, or more often about 10 mi. Amplitude (one-half of height from crest to trough) is largely a factor of terrain and is strongest about 3000 ft above the first crest. An amplitude of around 3000 ft is typical, and lift rates can exceed 1000 ft/min.

Larger wave occurs downwind from entire mountain ranges. The Sierra wave has already been alluded to, sometimes felt at altitudes above 50,000 ft. Weather satellites have observed waves extending over 500 mi downwind from the Rocky Mountains with wavelengths of about 25 mi. Lift in the up portions of these waves may exceed 2000 ft/min, and 1000 ft/min is normal. Similar large-scale waves have been observed by satellite east of the Andes Mountains in South America and over the ocean downwind from New Zealand. Wave probably forms downwind from the Brooks Range and the Fairweather Range (the most misnamed mountains in the world) in Alaska, but less often because of the wetter marine climate.

Although they are less prone to wave formation, the much lower Appalachian Mountains routinely produce wave extending over 100 mi downwind with wavelengths of 10 mi. Glider flights in wave east of the Appalachians from New Hampshire to Georgia have exceeded 20,000 ft in altitude. Lengthy periods of continuous lift in excess of 500 ft/min are very common. More is not unusual.

Flying in wave conditions?

Is it safe to fly in mountain wave conditions? Not always. Most texts will advise you simply not to fly in mountain wave, and at least in principle I agree. But to a degree, this question has already answered itself. Otherwise how would we be able to draw on so much experience from people who have flown in wave? Flight Service may very well be able to tell you when wave conditions are developing. You can probably still fly even in a small airplane, but you have to take certain precautions. Flying in the wrong part of wave can be a miserable experience. Flying the same day in a small airplane at altitudes below where the waves have formed can be extremely turbulent and even dangerous. High winds can mean the conditions below the altitudes where the waves are forming can be too strong for many small aircraft.

Your route of flight will definitely have to avoid the downwind side of high ridges and mountain ranges. This is easy if your route of flight is north or south—just stay west of all the ranges. It is much harder if you have to go east or west. If you have any doubt whatsoever, consult with local pilots who have flown similar aircraft over the route in similar conditions. It is very likely that the lee sides of ridges in these conditions have turbulence way beyond what your light four-seat aircraft can handle and sink way beyond what it can outclimb. If you have to cross one of these ridges, an altitude as much as 3000 ft higher than the ridge top might be necessary. In climbing out of the Denver area to cross the Front Range, that might mean an altitude of 17,000 ft, which is going to be beyond the capability of a lot of small aircraft. It might also mean a long climb through moderate turbulence and continuous sink in excess of 500 ft/min. You *could* go another day.

From a practical standpoint, on a day with strong wave conditions over the Rocky Mountains, there will also be strong downslope winds into the Denver area from the Front Range. Very small, low-powered aircraft simply will not be able to climb through this and get into wave higher up. Seemingly in contradiction with that, gliders launching east of Colorado Springs will be very capable of getting into wave and climbing to stratospheric heights. How? The gliders are being towed up another 10 to 20 mi farther east of the mountain front directly under the up portion of the wave. After bumping through some turbulence, up they go. What they cannot do is to penetrate the high winds aloft and to work their way west over the mountains.

Very severe turbulence can always be expected and sometimes be encountered below the lowest lenticular clouds in a stack. Go well around those areas. If the wave amplitude (wave height) becomes high enough and the climbing and descending faces of the wave are steep enough, a sort of cutoff eddy can form underneath the wave crests—in effect, a rotor not directly associated with a ridge. This sort of turbulence can be extreme. Here is a cautionary example. In the early 1980s, turbulence of this type was implicated in the crash of a Boeing 737 near Colorado Springs, Colorado. (Later the cause was thought to involve the rudder.) Airline pilots flying the 737 simulator into these conditions were frequently rolled inverted *even though they knew what was coming.*

For an inexperienced pilot, or even an experienced one in a small plane, one of the most dangerous aspects of such a flight can be the very prolonged periods of very strong sink (or lift) encountered when trying to cross the mountains from the downwind side. If you are in strong sink, get out of it. It helps to envision the probable orientation of the wave. It will be parallel to the range of mountains upwind of you. Flying perpendicular to the mountain range either toward or, better, away from it will get you out of sink. Flying parallel to the mountain range while in sink will just keep you in sink. You are not flying out of it. Since the sink you encounter can be far stronger than your ability to climb, approach mountain ranges at about a 45° angle. That way, if you have to turn back, it will be an easy turn. You could consider flying back away from the mountains until you find lift, climbing in the lift band, and then trying to cross the mountains again with greater altitude. Remember, though, if you had this much trouble crossing the first ridge or first range of mountains, you are due for more of the same as you progress farther into the mountains. You might want to try again early the next morning. In a turbocharged aircraft with a high enough service ceiling, you can climb above the worst of the wave almost all the time, given enough room to climb over the first wave you encounter.

The greatest difficulty in a small plane is fighting your way upwind over the mountains. Just briefly consider the other situation—a hypothetical encounter with strong mountain wave when headed downwind and crossing the mountains at less than twice the mountaintop elevations: Getting over is not going to be a big problem, as you will fly into strong lift on the upwind side. But now as you cross at an airspeed of 190 knots in your turbocharged Mooney, your ground speed is that plus the wind speed of 40, or 230 knots. At 230 knots you bounce uncomfortably through the very strong, abrupt turbulence on the lee side of the mountains. Then almost instantly you slam into an unexpected climb of 2000 ft/min (inadvertently busting through 18,000 ft and class A airspace in the process and getting the unwanted attention of the en route instrument controllers). The g forces are pushing your stomach down into your abdominal cavity, but not for long. Through the crest of the wave and whammo! Everything loose in the cabin is bouncing off the roof, the turbulence is threatening to rip the yoke out of your arms, and the ground may be very far below you, but it is headed up fast. And this experience doesn't just end. There could well be five or six more waves to go through before it is over. Slowing way down might help. Dropping the landing gear might help you do that, depending on both your exact situation and the speeds at which the gear on your plane can be operated. Some people also advocate pulling the throttle way back to slow down the climb, although with a turbocharged engine, rapidly pulling back the throttle can shock-cool the engine and crack cylinders. If you are flying on autopilot, definitely turn it off right now. The autopilot will try to dive down to counter the lift until you are way above maneuvering speed, and then try to climb through the sink until it stalls. Riding out lift, sink, and turbulence of this magnitude while flying in a normal attitude is the most comfortable thing for both your passengers and your airframe. If you could do it, climbing a couple of thousand feet higher would still have you in very strong up- and downdrafts, but less turbulence because you would be in the more laminar flow of the wave rather than the turbulent flow beneath it. Was it really necessary to be making this trip right now? Don't you wish you had delayed it until very early tomorrow morning?

In September 2001, flying west out of Denver at an assigned instrument altitude of 16,000 ft, I encountered the up part of mountain wave over South Park, east of the Collegiate Mountains. The result was an unintended climb through 21,000 ft. Of course we alerted the controller to our predicament and were cleared through those altitudes, and with a pressurized cabin there was no problem. The air was much smoother up higher, and we elected to stay up. Without pressurization or oxygen, there would have been a real danger of hypoxia, so it would have been necessary to lower the landing gear, deploy speed brakes, and reduce power to idle. While those steps would have produced a more than 1000 ft/min descent to counter the 2000 ft/min climb, the result would still have been an unintended 1000 ft/min climb well above 18,000 ft msl. Staying lower would have put us directly in the lee of the Collegiate Mountains in rotor turbulence. In a very small, low-powered basic aircraft, this would have been a very dangerous predicament except for one thing: A very small basic plane could not have dealt with the sink and head winds necessary to get up out of the Denver basin and up into the wave in the first place.

As with any mountain flying, cultivate a willingness to turn back. In wave conditions, you may have strong winds, or you may have very mild wind and turbulence on the ground. On your initial climb-out you will gain a pretty good feel for what you are flying into. If you don't like it, turn around. Go back.

Nacreous clouds

Related to wave phenomena are *nacreous clouds*. These are extremely high, sometimes above 100,000 ft, very thin clouds of ice crystal. The name refers to their mother of pearl coloration. These represent the entire atmosphere being lifted, usually in passing over a mountain range. This gives you an idea of just what a large-scale mountain wave very occasionally has. As this is being written, a Swedish team is designing and constructing a special glider intended to use these conditions to set a glider altitude record of 100,000 ft over New Zealand.

Kelvin-Helmholtz wavelets

Kelvin-Helmholtz wavelets form a very short-duration pattern of little wavelets in the shear layer between two air masses. They are an indicator of localized turbulence. Have I spelled this correctly? The reason I don't know is that the Kelvin-Helmholtz cloud is not an officially recognized type, but you will see them nonetheless. I am indebted to a seminar conducted by Margaret Lamb, Sue Baker, and David Whiteman for a discussion of these clouds. These people conduct seminars almost every year, and I intend to attend any of them I can. So should you.

The way to learn when these seminars are held is to join the Colorado Pilots Association, whose address is found at the end of the book.

Seen from the side, Kelvin-Helmholtz (K-H) clouds resemble a cartoon depiction of ocean waves. From below, they can resemble a closely packed series of small billows, looking like the ripple pattern of sand on the beach. K-H clouds can persist for a few minutes, but in general they are an extremely transient phenomenon. They may last only a few seconds, but are likely to reappear several times. Photographing these is like getting a picture of an unidentified flying object. You swear you see it. You rush into the house to grab a camera; but when you come back outside, it is gone. When seen from the side, Kelvin-Helmholtz wavelets are readily discernible as being in very rapid rolling motion. Sometimes instead of a series of K-H wavelets, the same conditions produce a single little rolling *billow* cloud, which is in effect a single K-H wavelet. When seen from that angle, it is very apparent what is going on. One air mass is moving rapidly over another, and the K-H clouds are a series of vertical eddies along the interface. In idealized form, the air masses rolling over each other form a series of horizontal rotating cylinders of air. K-H clouds form in the low pressure of these cylinders. While these clouds can be large, they are typically small, with vertical dimensions often much less than 500 ft. If you see a cloud like this in motion, you don't need to be told that it marks turbulence. How much

Kelvin-Helmholtz Wavelets and the Wind Mechanism Producing Them

The mechanism is a stronger wind overriding a less strong one. A series of horizontal rolling cylinders form in the shear layer between the two air masses. Lower air pressure of the faster moving air produces clouds in the characteristic wavelet form.

Seen from the side, these resemble something like a classic cartoon depiction of waves on the ocean, or perhaps something which might be painted in bright colors on the side of a 1950s hot rod.

This is a very short-lived, though usually recurring phenomena. Getting a picture of them is like photographing a UFO — you swear you saw it but when you rushed back outside with your camera, it was gone.

The Kelvin-Helmholtz cloud is a marker of shear layer turbulence. The photos were all taken off my deck in Telluride. In the photo, the wavelets are forming on top of a small, short-lived cumulus cloud. They can also sometimes form without an associated cloud, and the shear layer turbulence need not form any cloud if the air is dry or the temperature-dew point spread great.

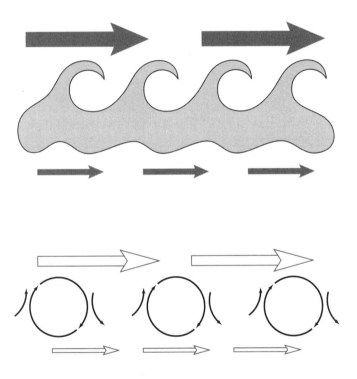

turbulence? Over how great an area? Those questions are unanswerable, but you are well advised to fly above or below rather than right at the level of these clouds. Anecdotally, turbulence is probably more extensive downwind than upwind of these clouds.

The smoke rising from a cigarette in a still room provides an illustration of Kelvin-Helmholtz waves and provides something else for you to do other than be annoyed by the smoke. Initially the smoke rises in a smooth, laminar column. As it rises faster, it produces a series of smooth, tight, little rolling eddies on both sides of the ris-

Kelvin-Helmholtz wave clouds. Getting a picture of these is like photographing a UFO—you swear you saw it, but when you grab your camera, it is gone. These clouds mark a shear layer between two air masses. They are of very short duration—sometimes only a few seconds. The rolling motion is obvious to the naked eye and clearly indicates turbulence to an experienced observer. On this day winds aloft were about 50 knots, and the area was covered by a severe turbulence sigmet. Compare with illustration on page 122. Fletcher Anderson photo.

ing plume. Still higher, it is rising so fast the airflow becomes turbulent and the smoke dissipates throughout the room.

Small-scale mountain winds

As with clouds, small-scale localized wind phenomena are of great concern to the pilot of a small plane. Sometimes these micro-scale weather events produce clouds. At other times they don't. They are sometimes useful and at other times are producers of unanticipated sink and turbulence.

Pilots tend to be aware of and take appropriate actions with regard to larger-scale weather phenomena. Small-scale weather phenomena, extending only a few miles or less and involving only a few thousand feet of elevation, are not always predictable even to experienced pilots. They are the result of sometimes very complex causes whose presence cannot always be detected from the air. They can

This cloud looks like a bunny. Fletcher Anderson photo.

be encountered very abruptly with little warning. For these reasons they can pose a greater hazard than much larger, stronger, but more easily observable large-scale weather.

Overall, regional winds aloft can be obtained from Flight Service. (Have you memorized their phone number, 1-800-WX BRIEF, yet?) While these observations and predictions give you an overall framework for the area, it cannot be stated too emphatically that the actual wind directions and velocities you encounter in the vicinity of the mountains, particularly when flying lower than ridge top elevations, will not be the same as forecast winds aloft. Some of the small-scale mountain-generated winds are as follows:

Katabatic/anabatic winds

Warm air is less dense and rises; cold air is denser and sinks. During the day the sun warms the ground, which in turn heats the air by convection. The warm air rises. After about 10:00 a.m. on a typical summer day in the mountains, this warm air starts flowing up the sunny mountainsides. As that flow develops, still more warmed air begins flowing up the valleys. This is *anabatic wind*.

In late afternoon, the angle of the sun's rays is too close to parallel with the ground to have the same heating effect. Convective heating

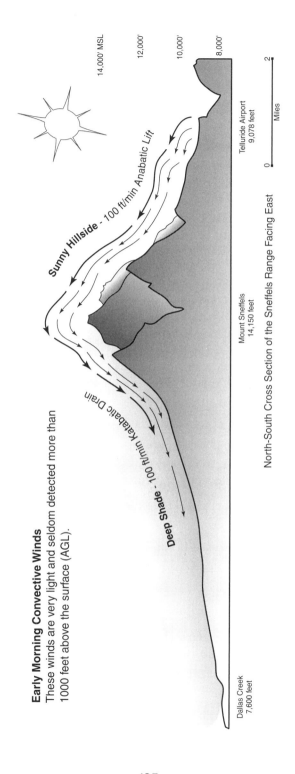

Early Morning Convective Winds

These winds are very light and seldom detected more than 1000 feet above the surface (AGL).

Sunny Hillside - 100 ft/min Anabatic Lift

Deep Shade - 100 ft/min Katabatic Drain

14,000' MSL

12,000'

10,000'

8,000'

Telluride Airport
9,078 feet

Mount Sneffels
14,150 feet

Dallas Creek
7,600 feet

North-South Cross Section of the Sneffels Range Facing East

0 — Miles — 2

and adiabatic flow die down. Around sunset, the stored heat in the ground on the mountainsides has been or will soon be radiated away. Cooler air now begins flowing down the mountains and collects to flow down the valleys as a river does. This is *katabatic* wind.

You can think of the daily katabatic/anabatic valley winds as the valley's daily breathing. Valleys of any size will have these winds. Wind velocity, among other things, is a factor of the area being drained and the rate of heating and cooling. Winter may exhibit strong valley-draining katabatic winds, but lighter anabatic valley upflow because snow-covered ground reflects sunlight back into space rather than absorbs it and produces convective heating. Dry ground in summer heats up faster and the sun's rays are more vertical, so stronger anabatic winds are the norm. Most mountain valleys will have an early morning katabatic and midday anabatic flow of around 4 knots, plus or minus a large margin. The Jordan River Valley, connecting Utah Lake to the south with the Great Salt Lake to the north, has very little elevation change, but drains one very large basin into another. In between the two lakes, the valley is constricted by a low ridge called Point of the Mountain. Predictable 10-knot daily valley winds over this ridge make Point of the Mountain a world-renowned ridge as a soaring and paragliding Mecca.

Elk hunters who understand this phenomenon are much more likely to get an elk. During the predawn hours they climb up very high in the mountains so that they can hunt downward from a ridge. Not only are the elk moving uphill toward them after feeding all night, but also the anabatic wind is bringing the smell of the elk up to the hunter and taking the hunter's scent away from the elk. In the evening, the same hunter will hunt from the valley when katabatic valley drain will again take the human scent away from the elk.

Temperature inversions

Normally temperature decreases as you go up in elevation. A temperature inversion exists whenever the rate of cooling as you go up is less than the standard rate. In mountain valleys, the temperature on the valley floor early in the morning can actually be substantially colder than the temperature only 1000 ft higher. These temperature inversions have already been referred to because of the very low-lying stratus clouds associated with them. Particularly on a cold, calm, very clear night, as heat radiates off into space, cold air flows downhill and settles in the valley bottom. By morning, a very stable layer of cold air anywhere from a couple of hundred up to over 1000 ft thick fills the valley. Particularly in winter when snow reflects away the sun's heat, it may take a lot to dislodge one of these stable inversion layers, and they may persist all day long. Regardless of very strong winds aloft, it is usually very possible to fly in relatively calm conditions early in the morning if you stay well below the ridge height.

Thermals

Thermals have been mentioned in connection with cumulus cloud formation. If there is insufficient moisture in the air to form clouds, the sun's heating of the ground may be even stronger, and convective heating will produce even stronger thermals.

On a day with high barometric pressure and stable air, it takes a lot of energy to get a thermal to release from the ground. Once it does release, the resulting thermal tends to form a very compact, fast-moving bullet of air. Thus, mid-afternoon convective turbulence on one of these days becomes a very uncomfortable, but not necessarily dangerous, ride. Flying higher generally gets you out of this turbulence. The ride down to landing can be very rough, but right down at ground level things may smooth out enough to make the landing uneventful.

On a day with low barometric pressure and unstable air, the climb rate in thermals may be stronger, but the thermals themselves require less energy to release from the ground. Instead of compact bullets, you find broad rising columns of air, and paradoxically the ride may be gentler up until the time afternoon thundershowers develop. These are the days glider pilots live for.

Thermals have a tendency to stick to the ground. Just as water dripping off a faucet will run along your finger to your fingertip before it drips off, a thermal will tend to track up the valley wall before finally breaking off at the ridge top. For that reason, it is relatively easy to find thermals by cruising along the top of a ridge, while it is less likely to find them over the middle of a valley unless the valley is fairly wide.

The acceleration of air up the sunny side of a ridge can sometimes increase the wind speed over the ridge top and increase both rotor and ridge lift.

Ridge lift/rotor

Ridge lift and rotor are opposite sides of the same coin. They are the results of wind blowing over a ridge. Let us examine the winning side

Rotor. The blowing snow off the ridge in the background is an indicator of rotor, but this pilot is either too far downwind to feel it, being thrashed by the rotor, or enjoying himself so immensely doing rolls in formation with the camera plane that he hasn't noticed. Take your pick! John Kounis photo.

of the coin toss first. When wind blows against a ridge, the air is compressed (reducing turbulence) and deflected upward, producing ridge lift. Glider flights of over 1000 mi have been made as long ago as the early 1960s by flying the upwind side of ridges from upper New York State all the way down the length of the Appalachian Mountains deep into Georgia, never getting more than 500 ft off the ground. The Golden Trough in the midst of the mountains of British Columbia produces similar flights of far greater altitude when convective lift augments ridge lift.

The problem for aviators is the downwind side of the ridge. Some quick calculations will show us the beginning of the problem: 1 mi/h = 88 ft/min, and 2 mi/h = 176 ft/min.

Now picture yourself sitting on a ridge high up in the mountains above timberline. It is about 9:00 a.m., a calm, clear, beautiful morning. But it is not quite absolutely calm. There is a just barely perceptible hint of a breeze wafting up the sunny side of the ridge—a breeze so light you almost have to wet your finger to tell which way it is blowing. That is a 2-knot wind. That small amount of wind is already creating close to 100 ft/min of lift out in front of the upwind side of the ridge. Birds are soaring. If by chance the imaginary ridge you climbed were Aspen Mountain so that you could pretend to get up there riding the Gondola, then you are noticing a few paragliders already floating back and forth like dandelion seeds above their launch point in this light lift.

On the shady, downwind side of the ridge, having been lifted as it crossed the ridge, the gentle breeze is now sinking back down at about 176 ft/min. The wind is not blowing straight down, but perhaps a sink rate of 100 ft/min is already possible. Get out your aircraft operator's manual. If you are flying a Cessna 172, a Piper Cherokee/Arrow, a Beech Skipper, or any other small basic aircraft, depending on the temperature and the exact altitude of the ridge (timberline in Colorado is about 12,000 ft msl), that 100 ft/min sink may be very close to, if not more than, your best rate of climb at full gross weight. You could already be losing altitude in trying to fly across this ridge.

Across the tops of the mountains 15 knots of wind would not be unusual. On the *upwind* side of the ridge, the air might be a little choppy and the ridge lift quite strong, but the flying would not be uncomfortable.

The downwind side of the ridge is now nowhere to be in a small air-craft. Overall, you could expect to be encountering sink rates of close to 1000 ft/min as the air comes down the back side of the ridge. Your small basic aircraft cannot possibly outclimb that sort of downdraft. Worse, the wind cannot maintain a nice smooth laminar flow over the ridge crest and down the back when the wind is this strong. It will set up a sort of vertical eddy called a *rotor*. See a simple diagram of what this looks like. The rotor in reality will not be quite the nice smooth curl of the diagram. It will be a tumbling mass of turbulence which may extend a considerable distance downwind. How far downwind? The only way to know is to fly into it—the distance will vary greatly with temperature, humidity, air stability, terrain, and wind velocity. Rather than deliberately fly right into known sink and turbulence, you might prefer to stay on the upwind side of the ridge.

Venturi

In all probability, if the prevailing wind is blowing 15 knots, then the wind over a ridge top perpendicular to the wind is at least 20 knots or above. So is the wind through a mountain pass. So might be the wind up or down a narrow valley. So will be the wind wrapping around the shoulder of a mountain. Why?

There is not an infinite amount of space through which the air can move. Air blowing against a mountain is forced around the sides. It cannot simply spread out because more air is already occupying the space to the sides. The air is compressed and slowed. Then after it squeezes past the sides of the mountain, it expands and is acceler-ated. Air blowing over a ridge is compressed and lifted on the up-wind side, but is not lifted all the way up into space. It, too, squeezes in front of the ridge and is slowed, then forced over the top and ex-pands and acquires a higher velocity. Air blowing around a mountain is squeezed through the pass. Forcing the air through the pass in-creases its velocity on the downwind side.

These phenomena are milder with unstable air when the air *can* be forced upward a long way. They become much stronger with stabile air and high pressure. Although a meteorologist might cringe at this analogy, think of high pressure as a roof holding down the atmos-phere. Unable to go up, the wind must find its way around any obstructions it encounters.

At times winds that more than double the velocity of the predicted prevailing wind may be encountered in narrow mountain passes or over sharp mountain ridges. Concurrent with these doubly strong winds, you will encounter doubly strong sink and turbulence on the lee side of the pass or ridge.

Monarch Pass, Haggerman Pass, Independence Pass, and Gore Pass in Colorado, to name just a few examples from a very large group, are very prone to developing these conditions.

Anomalous reverse-direction winds

Here is an example of one of those situations where you *think* you know what is going on, but you should not bet your life on it. You are flying up a mountain valley, intending to cross the pass at the head of the valley. Your route of flight is east, with the wind out of the west. You are climbing in light lift, which only serves to increase your confidence about clearing the pass. All your mountain experience tells you that you are on the "good" side of the pass, with all the micro-scale weather you can detect working to help you. If it isn't an absolutely perfect day, it is at least good enough for the people you fly with. Your biggest concern is the quality of the scenery.

Now things subtly change. There is the least little burble of turbulence. But hey, you are in the mountains, aren't you? Then you are no longer climbing. Then you are actually losing altitude, and you are no longer going to be high enough to clear the pass. The transition from a slight climb to a very extreme descent could take as little as 2 s.

(Where did I get that figure? As I write this, I was able to demonstrate a transition from a 500 ft/min climb to a 1500 ft/min descent to an instructor from Florida in less than 2 s on a mountain orientation flight in the Weminuche Mountains this very morning. Today was a very mild day with almost no convective activity.) What phenomenon was directly responsible for this wind? I don't know, but I was already prepared to turn around when we hit it.

This could be a very serious situation indeed, particularly if the transition is abrupt and the sink is strong. What you have encountered is an anomalous wind blowing over the pass in the opposite direction from the prevailing winds you encountered flying up the valley. Let us examine four different conditions that can produce this phenomenon, while agreeing that there are no doubt other circumstances we cannot detect which might also do something similar.

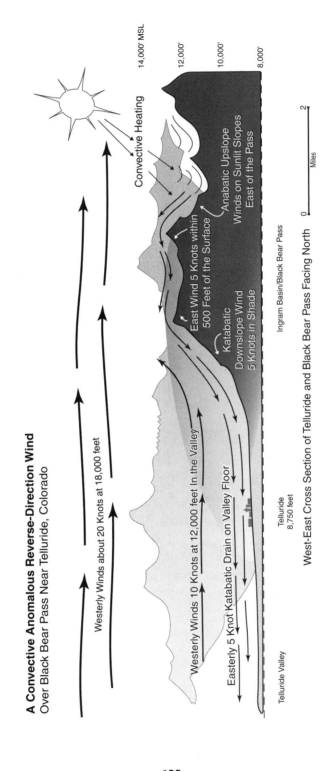

A Convective Anomalous Reverse-Direction Wind
Over Black Bear Pass Near Telluride, Colorado

Westerly Winds about 20 Knots at 18,000 feet

Convective Heating

Anabatic Upslope Winds on Sunlit Slopes East of the Pass

East Wind 5 Knots within 500 Feet of the Surface

Katabatic Downslope Wind 5 Knots in Shade

Westerly Winds 10 Knots at 12,000 feet In the Valley

Easterly 5 Knot Katabatic Drain on Valley Floor

14,000' MSL

12,000'

10,000'

8,000'

Telluride Valley

Telluride
8,750 feet

Ingram Basin/Black Bear Pass

0   2
Miles

West-East Cross Section of Telluride and Black Bear Pass Facing North

Reverse-direction convective wind

Early in the morning, the wind might be out of the west, but the *sun* is in the east. The prevailing wind will blow up a west-facing valley, but the head end of the valley will be in deep shade. On the other side of the pass, the east-facing slopes are soaking up sunlight. The prevailing wind blowing up the valley should produce lift near the head end of the valley, and wind velocity ought to be accelerated as the wind venturis through the pass—but it doesn't. Instead, the sunny east-facing slopes east of the pass are creating a convective anabatic upslope wind. The deeply shaded west-facing slopes west of the pass are still very cool and are producing a downslope draining katabatic wind. While the wind 2000 ft or so above the pass is westerly, the wind at ground level up to perhaps 500 ft above the pass is flowing in the opposite direction because the upslope anabatic wind is draining over to the downslope katabatic wind. There is a shear layer where the two winds flow over each other, which will produce a little turbulence, and a head wind and downdraft near the ground just where you were anticipating a tailwind and lift.

Most of the time, this particular reverse-direction wind is relatively light. In another hour with the sun reaching the shady side of the pass, the reverse flow will stop. For half an hour or so, the sunny-side convective upslope wind will meet the prevailing west wind right at the top of the pass. If the air is unstable, the result will be lift on both sides of the pass converging to form a lift band extending a couple of thousand feet up. Then the west side of the pass will have both sun and prevailing wind, and the anomalous conditions will disappear.

Reverse-direction deflected winds

Rather than delve too deeply into matters hypothetical, let me just describe two examples. Flying directly from Telluride, Colorado, to Durango, Colorado, you would head south over Lizard Head Pass. This is a low pass (10,500 ft) scarcely above Telluride's pattern altitude. Telluride is at the very head of a steep-walled box canyon; the airport is 4 mi to the west farther down the valley. Lizard Head Pass is about 5 mi south of the airport, running north-south into the next valley. A west wind at the airport should be expected to fill up the head end of the valley and flow out to the south over Lizard Head Pass, the lowest pass out of the valley.

What actually happens a lot of the time is quite different. The next valley to the south on the other side of the pass is also open to the

Telluride looking east. This valley often experiences a reverse-direction convective wind in the morning. By afternoon when this photograph was taken, the upper end of the valley typically sees both ridge lift from the prevailing west wind and convective lift as the steep west-facing slopes are heated by the sun. Notice the small puffy cumulus clouds marking thermals at the head of the valley. Ron Kanter photo.

west and is far larger than the narrow Telluride box canyon. A much greater volume of air is blowing into that larger valley and is *deflected north* into the Telluride Valley.

Glenwood Springs is at the very lower end of the north-flowing Roaring Fork River Valley. The valley joins the deep and narrow west-flowing Colorado River Canyon at the town of Glenwood Springs. The airport is just south of town. A strong southwest wind ought to blow down the valley, producing a very strong headwind for runway 14. It often doesn't. Instead, the wind is so strong up the larger Colorado River Valley that when that up-valley wind is constricted by narrow Glenwood Canyon, it wraps around the corner and up the Roaring Fork River Valley. The actual surface wind in the bottom of the valley up to about 2000 ft agl is blowing north, in the *opposite direction to the southwest winds aloft*. On those days, the descent into the narrow canyon at Glenwood as well as the climb-out can be miserable experiences with constant multiple wind shifts.

Similar situations can be encountered all over the mountains as the wind wraps around the corner out of a larger valley into a smaller one or out of a valley perfectly aligned with the wind into one angled across the wind.

Reverse-direction pressure drainage winds

The existence of these winds was discovered very recently. In 1990 G. L. Stone and D. E. Hoard published "An Anomalous Wind between Valleys—Its Characteristics and a Proposed Explanation," in *Preprints of the Fifth Conference on Mountain Meteorology,* Boulder, Colorado. Stone and Hoard proposed uneven heating as the cause of a reverse-flow wind between Utah Lake Valley and the Rush Valley 37 mi to the west.

In their study of airplane crashes in the Aspen, Colorado, area, Sue Baker and Margaret Lamb discovered four accidents in particular which displayed symptoms of anomalous reverse-flow winds. Initially they looked for thermally driven reverse flow, but ultimately they found strong evidence for a previously unreported phenomenon— *pressure drainage reverse flow.*

Air flows from areas of high pressure to areas of low pressure. Given the prevailing westerly flow of air across the North American continent, you typically encounter a pressure gradient of higher-pressure air west of the Rockies and lower-pressure air east of the Rockies. The prevailing winds are dammed up by the mountains, and the result is a compression zone upwind and a lower-pressure zone immediately downwind of the mountains. On the dates of the crashes studied, prevailing winds were westerly, but the pressure in Leadville, east of Haggarman and Independence Passes on the Continental Divide, was *higher* than in Aspen, west of the passes—the reverse of normal conditions.

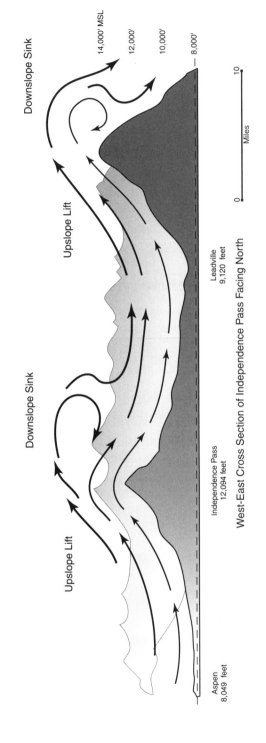

Normal West Wind Over Independence Pass

Downslope Sink

14,000' MSL

12,000'

10,000'

8,000'

Upslope Lift

Downslope Sink

Leadville
9,120 feet

0   10
Miles

Independence Pass
12,094 feet

West-East Cross Section of Independence Pass Facing North

Upslope Lift

Aspen
8,049 feet

136

Reverse-Direction Pressure Drainage Wind

The Leadville barometric pressure is 0.9 to 2.0 millibars higher than Aspen. An anomalous reverse-direction wind of as much as 10 knots blows at the surface over the pass from the East against a 20-knot prevailing wind. Several small aircraft headed east from Aspen have crashed in these winds!

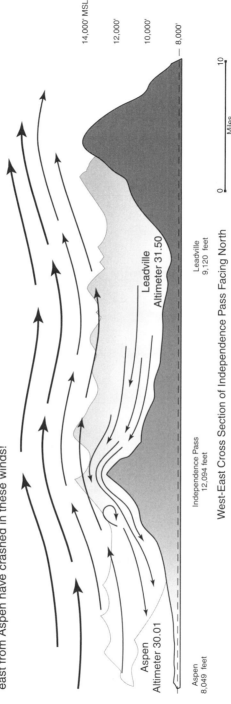

West-East Cross Section of Independence Pass Facing North

Small aircraft climbing in light lift and west winds west of the passes very suddenly encountered wind shear in some cases abrupt enough to cause a stall, and then downdrafts all the way to the ground, taking them down at an observed 2000 ft/min. By the way, these were experienced mountain pilots, some of them flying familiar aircraft over familiar routes. The extreme abruptness of the episodes was a major contributing factor in the accidents.

The very broad Arkansas River Valley just east of the Continental Divide may make Haggarman and Independence Passes unusually prone to this particular kind of pressure drainage reverse-flow wind, but no doubt it can be encountered elsewhere. Three crashes in the mid-1990s on the northwest side of the Uncompahgre Plateau south of Montrose, Colorado, occurred on days when the pressure gradient between Montrose to the north and Telluride to the south exceeded 20 mbar, or 0.2 in of pressure. The local topography can sometimes produce a more than 50-mbar, or 0.5-in pressure difference between Telluride and Montrose, only 25 nautical mi away. This is the equivalent of a 500-ft pressure altitude difference, producing the equivalent of a waterfall of air that high flowing toward the low side. Clark King has measured similar reverse flows over Corona Pass on the divide west of Boulder, Colorado, publishing his results in 1997. David Whiteman has been studying the central Rocky Mountains and observed numerous similar situations.

Typically, these anomalous pressure drainage reverse-direction winds are only encountered in the lowest 1000 ft of elevation over the pass. Flying higher gets you above them.

I stress that this is a newly discovered phenomenon and it is not yet entirely understood. Unusual pressure gradients sometimes produce pressure gradient drainage winds, but not always. Being aware of the possibility makes pilots much more prepared for the encounter. By publishing reports and presenting this information in numerous seminars, Baker and Lamb have saved an unknowable but probably large number of lives.

Microbursts and gust fronts over ridges and passes

Whatever the prevailing winds might be, the very strong downdrafts under a cloud associated with a microburst can produce a very strong gust front blowing directly away from the place where the microburst hit the ground. In the rugged terrain of the mountains, these gust fronts are channeled up or down the valleys and can run for several miles. They can and occasionally do blow over a low pass or ridge and drop down into the next valley. As they do, there is a short-lived very strong wind with associated rotor turbulence on the lee side. This can easily be strong enough to blow in the opposite direction and underneath the prevailing wind.

The visual clue that something like this might be possible is the presence of virga underneath clouds. The source of the gust front is the downdraft that produced the virga.

This anomalous wind is of particular personal significance to me. In 1991 I was caught by such a reverse-direction gust front while flying my paraglider in what was light upslope lift north of Sunlight, Colorado, and tumbled through the lines. I spent 3 months in the hospital and 1 year in rehabilitation. Eventually I decided to stay with conventional aircraft, and I paid much more attention to virga.

Summary of anomolous reverse-flow winds over passes

For a pilot, the important fact to remember about these winds is that they are very real. They occur much more frequently than is generally appreciated. Their causes are myriad and not entirely understood. It is beyond reason to expect a pilot in flight to analyze all the available clues that these winds might be likely. *As a rule, these winds are encountered within 2000 or 3000 ft of the ground, often less than 1000 ft above the ground. A pilot is at greatest risk from these winds when climbing toward a pass at or below pass elevation.* Being prepared for these winds even when they are not expected is the only defense.

Have you had enough of weather? Are you ready to move on to flying?

4

Mountain flying strategy

Flying the mountain environment

If you have finished wading your way through the last few chapters, then you should be feeling very cautious about flying in the mountains, and rightly so.

You will be flying at the upper limits of the atmosphere, the weather, the terrain, your aircraft, its engine, and your body. You will be flying in a realm where the weather *routinely* exceeds the operating limitations of your aircraft. You will be in a realm where that weather is rapidly changing and inherently unpredictable.

In this environment, a larger, more powerful, higher-performance aircraft may raise the limits, but the limits of *any* small aircraft can be exceeded.

In this environment, superior flying skills may raise your personal limits, but regardless of your experience and ability, those limits can be exceeded.

Because of these factors, you are flying in an environment where the exercise of good judgment is far more important than skill or equipment. The exercise of good judgment and appropriate caution is your primary safety tool in the mountains. While this is a chapter about how to fly in the mountains, it could also be read as a chapter on when and why not to fly.

It would be perfectly logical just to decide never to fly in the mountains at all. *Nevertheless, on a perfect day, flight over the mountains in a small aircraft is safe and easy and requires no more skill than any other flying.*

Good planning can mean you are in the air only on perfect days. Rapidly changing mountain weather is the reason you need to know and practice strategies and techniques for mountain safety. You don't need to know these things because you plan to go on bad days. You will still be able to go flying—frequently, safely, and pleasantly! You just have to be a little more aware than you need to be in flat country.

For convenience sake, this chapter is broken down into various phases of the flight.

Time of day

Desert in early morning. Notice that this pilot is flying the sunny, lifting side of the ridge, but mostly notice how serene the desert is at 6:00 a.m. John Kounis photo.

In any small aircraft, the time of day you select to cross high mountains can be particularly important. If you want the simplest rule possible for any very light airplane, it is this: Finish the flight by noon, and stay on the ground.

A typical summer day in a typical mountain valley begins with cool air temperatures and virtually calm winds or a very slight valley drain. Soon, however, the heat of the sun begins to produce convective activity, and by midday the level of convective turbulence can be greater than most people enjoy dealing with in a small aircraft. Given any moisture in the air, convective buildup will lead to afternoon thunderstorm development. The turbulence over the mountains might only be uncomfortable, but the strength of downdrafts could make it impossible to climb over some passes, and strong, gusty winds in the valley bottoms near your destination airport could make landing dangerous. While some of this weather is the result of prevailing winds, most of it is heat-driven, and around 5:00 or 6:00 p.m. when the angle of the sun is low, there is no longer enough heat being produced to drive the system. Things will smooth out considerably, and in Colorado in the summer, it is still light until around 9:00 p.m. Departing so early in the morning that you can just barely see the runway will make for a much safer and more comfortable ride. Returning as late as feasible will do the same.

Desert afternoon. Summer midday desert convective activity virtually always produces light to moderate turbulence, which just as predictably dies down late in the day as the sun's rays hit the earth at a lower angle. John Kounis photo.

Time of year

Winter winds aloft in the midlatitudes are far higher than summer winds, but paradoxically some winter flying can be far less demanding. The air is colder and therefore denser, so you have greater power and greater lift. Although winds can be strong, unless you are in the midst of a storm, cold, clear nights produce very strong temperature inversions which can persist well into midday. Despite strong winds aloft, the air under the inversion can be very stable. You can sometimes fly all day long down in the valleys with no turbulence at all. As in summer, a very early departure can allow you to make both the takeoff and the landing in still air under the inversion despite winds aloft too strong to fly comfortably in.

There is an exception to this, though: The very high peaks of the continental divide push right up into these higher winds aloft in winter. One of the photographs in this book shows 70-knot surface winds scouring the Collegiate Mountains in Colorado, as viewed from a Cessna P210 at 17,500-ft elevation flying in winds around 85 knot. While flying in the valleys in the morning can be very pleasant, crossing the continental divide in a small plane in winter can be a little more adventure than most of us look forward to. The Colorado Pilots Association policy is not to offer its highly regarded mountain flying course before June 1 or after September 15 each year, to avoid trying to cross the continental divide in these conditions. This is not to say that optimal conditions cannot be found at any time of year; rather they are less likely.

Think behind

This sounds gimmicky, but it is the most important aspect of all mountain flying, so here is a chance to read it again: *Think behind!*

The standard advice for all flying is to *think ahead*. Be ahead of the airplane. Anticipate what comes next and be ready. You know the drill. Lighthawk's check airman Dick Arnold stresses that you should do just the opposite: *Think behind!*

What does he mean by this? The likelihood in the mountains is that all sorts of things you intend to do might not happen. You might not be able to cross a pass. The weather might be turning bad before you can reach your destination. Especially in a small plane in marginal conditions, all sorts of things can happen that can oblige a change in plans. The standard mind-set of "this is my plan and this is going to happen next" has to change to "whatever I have planned isn't going to work out, so these are my options when it doesn't." How you will turn back is so much more important than what you intend that you should always be at least psychologically surprised when things actually do go the way you planned.

Takeoffs

Air being less dense at higher altitude, true airspeeds required to produce lift being higher, and engine power being less, takeoff distances will be increased. Often mountain airports are built in areas where broad expanses of flat ground are not available. Particularly in smaller communities, the length of mountain runways can be rather short compared to the increased distances needed for takeoff. Particularly in summer, especially when you are carrying a heavy load, it is important to determine whether you will be able to take off within the runway distance available. This distance can seem very long to someone accustomed to low-altitude operations.

As a very rough rule of thumb depending on the specific type of aircraft, at 8000 ft msl, your takeoff roll doubles while your rate of climb is around one-half that of sea level. This means that distances required to clear obstacles off the end of the runway can be almost infinitely long.

The Koch charts are reproduced as an illustration and are reasonably accurate most of the time for calculating takeoff and climb performance. Still better, the aircraft operator's manual contains takeoff distance charts, and these must be consulted! For most small aircraft, the takeoff performance charts stop at altitudes lower than those of many mountain airports. When you start extrapolating data from the charts to calculate your takeoff roll, be conservative and calculate climb performance as well!

Takeoff technique

Much has been written suggesting that various special, if not peculiar, techniques should be employed for a high-altitude takeoff. I will argue strongly that this is not the case at all. Because the air is thinner, you will have to be going faster over the ground to achieve the amount of lift needed for flight. However, your airspeed indicator also has thinner air going into it, so it is reading too slow. In fact, the airspeed indicator error is precisely equal to the extra speed needed for flight. *You should fly exactly the same indicated airspeeds at high altitudes that you do at low altitude.*

I have read articles advising you to use more flaps for a high-altitude takeoff. I have also tried this many times. For a low-powered plane, my advice is, Don't. Your engine is developing less power at altitude. Why would that make it more able to overcome the extra drag

Heavily loaded on a windless, hot summer day, this pilot needs all the short runway to get airborne. Notice that he is using 15° flap and letting the aircraft fly itself off in its normal climb attitude. John Kounis photo.

of more flaps? Use the *same* amount of flaps you would use for the type of takeoff you are doing at all altitudes. For a standard Cessna 172, this means zero flaps. For a 182, this could mean 20° flaps. The amount of flaps required for the shortest takeoff roll is aircraft-specific. If you are unsure, experiment on a very long runway.

There is no special high-altitude takeoff technique.

Accelerate down the runway with neutral pressure on the yoke. When you reach rotation speed, lift the nose to your normal climb attitude and let the aircraft fly itself off. If there is a crosswind, hold the upwind wing down with the ailerons. Stay on the ground until your speed is slightly higher, then pop it off a little more abruptly. If the field is soft, use more flaps *as specified in the operator's manual,* hold the nose-wheel just clear of the ground, lift off at low speed, and then accelerate in ground effect. Aren't these the same as the ways you would take off at sea level anyway? The only difference is that since your aircraft is flying with about one-half of its sea-level performance, more precise control of pitch attitude is needed to obtain optimum performance.

Accelerating to takeoff speed and then quickly pulling on 20° of flap is a technique favored by bush pilots. This works well for them be-

This is a very sound fundamental pilot technique for a soft field takeoff: The nosewheel is lifted off the ground as soon as possible, the plane becomes airborne at a very low speed and then accelerates to climb speed while the pilot holds it down in ground effect. John Kounis photo.

cause they have manual flaps which can be applied almost instantly. If you are flying an aircraft with electric flap controls, the slower rate at which the flaps deploy means this will be of little value, so you might as well begin the takeoff roll with the flap setting you need for takeoff. But if you are running out of runway or the trees at the end of the field are coming up fast, at the last moment you can drop in 10° or even 20° flap. This will balloon you over the obstacle, although your climb after you have ballooned over may be very poor and you may even drop back down. Opinion differs about adding more than 20° of flap. For the airfoils used on most small aircraft, adding flaps up to 20° adds both lift and drag. Adding more than 20° of flap increases drag further but adds no more lift.

Once you are over the obstacle, retract the flaps very slowly to avoid settling into the ground.

One possible exception to the last advice is the soft field takeoff. The soft field takeoff can be a real asset when departing a high-altitude runway, whether the field is actually soft or not. You had to practice this to get your private pilot's license. The idea is to take off with the nose held slightly high with the nosewheel just barely off the ground.

The aircraft will fly itself off the ground at less than its normal take-off speed. You then hold it down in ground effect until you gain enough speed to climb out. The ground roll for this type of takeoff is typically 10 percent shorter than that of a normal takeoff, depending on the type of aircraft. More importantly, you leave the ground at precisely the instant it becomes possible to do so. Unfortunately, in a low-powered aircraft, while this technique gives you a shorter ground roll, it *increases* the distance needed to clear an obstruction off the end of the runway. You find yourself airborne right at stalling speed, so even minor turbulence could be a problem. This is not necessarily the best idea for all conditions or circumstances.

Density altitude

On a hot summer day, *density altitude,* that is, the actual density of the air, can be quite a bit higher than mean sea-level altitude. Remember, the higher the altitude, the thinner the air? In conditions in which aircraft performance is already marginal, that extra degree of thinner air can make a takeoff impossible. Density altitude can be read off AWOS at larger airports that have an AWOS.

The FAA provides density altitude calculation charts, and one is copied here. Let us perform a sample calculation: The altitude at Leadville, Colorado, is 9927 ft above sea level, but the runway is 6400 ft long, which is makes takeoff possible, but long in almost any small production aircraft. Now, though, let's calculate the density altitude on a typical August afternoon. I advise you to do similar calculations from the chart for any airport you intend to visit.

Since it is a nice clear day with no storm in the vicinity, the barometric pressure is high. Let us use an altimeter setting of 30.2. This helps us—we can subtract 225 ft from field elevation to get a *pressure altitude* of 9702—not much difference, but it is a little lower. Next, though, assume a temperature of 78°F. This raises the *density altitude* to 13,500 ft—an altitude above the service ceiling of many small aircraft, an altitude at which they cannot even fly, much less take off.

Leadville abounds with stories of people who arrive in summer with a fully loaded small plane, top off with fuel, and then take off. They usually manage to get into ground effect, but cannot climb, and they drop out of sight off the end of the runway. This is not necessarily fatal if they took off to the south where the valley gradually drops away to lower altitude.

So for some small aircraft, Leadville doesn't work out very well, if at all, on a hot summer day, but then Leadville is the highest public use airport in the United States. Perhaps Aspen at an altitude of 7815 ft msl with a 7000-ft-long runway will work better. We are going to use a 1967 Cessna 172 operator's manual for these calculations because the Cessna 172 is the most widely flown general aviation aircraft in the world. An equivalent model Piper Cherokee may cruise a little faster, but needs noticeably more room to take off and has a slower climb.

At sea level with no wind, this aircraft at its 2300-lb gross weight takes off in 865 ft and clears a 50-ft obstacle in 1525 ft, i.e., in about one-fifth of the length of Aspen's runway, you would already be 50 ft in the air and climbing away at better than 600 ft/min. In Aspen on a 32°F day with the same load, it breaks ground in 1585 ft and clears a 50-ft obstacle in 3855 ft. This is a perfectly safe distance, but it is well over one-half of the runway rather than less than one-quarter, and the rate of climb right after takeoff will be about 300 ft/min. The aircraft is flying acceptably, but the takeoff distance about doubled; obstacle clearance distance more than doubled, and climb rate has been cut in half. Presumably you intend to travel to the west with this load, because in flying east across Independence Pass, your climb rate at 15,000 ft is only going to be 22 ft/min.

All this would work on a calm 32°F day, but Aspen sees summer temperatures as high as 90°F. For each 25° of temperature, increase the takeoff distance by approximately 10 percent. Takeoff ground roll is now 1878 ft, and obstacle clearance distance is now 4626 ft. Again, this is a rather leisurely takeoff, but certainly possible given the 7000-ft-long runway. Rate of climb decreases by about 20 ft/min for each 10° of temperature. Rate of climb right after takeoff is now down to only 120 ft/min. At the fence at the end of the 7000-ft-long runway, you would be scarcely 100 ft off the ground, if that. This is only just enough to clear nearby obstructions if you took off to the north. It is too slow a climb rate to either clear obstructions *or turn around* if you took off to the south. At somewhere around 12,000 ft, this airplane on this hot summer day will climb no higher.

Of course, there are Cessna 172s based at virtually all mountain airports, but unless they have had 180-hp engine conversions, they are generally flown as two-seaters most of the time. Even then, on a hot summer day, takeoff and climb performance is very marginal at mountain altitudes.

Best rate of climb really can be obtained only when you are flying straight ahead and all the lift in the wing is pulling straight up. In a bank, the lift is pulling in the direction in which the wing is banked. Peter Lert relates a story of spending 45 min buzzing Grand Lake Colorado in, of all things, an Ercoupe—one of the definitively low-performance planes of all time. He would fly as much of the length of the valley as he could, each time squeezing up a few more feet in altitude, but lose that slim gain each time he had to turn around.

Many aircraft of similar marginal climbing ability have taken off just fine from mountain airports on cold winter mornings, but bumped up against the temperature inversion layer and climbed no more. Once the aircraft is above the top of the inversion layer, the slightly warmer air is just enough thinner to stop the climb.

Even such a basic thing as using enough right rudder to counter propeller torque when flying in a nose-high attitude can make the difference between actually gaining altitude and not.

Uphill/downhill takeoffs

Flat ground being at a premium in the mountains, it should come as no surprise to learn that many mountain runways are sloped.

Takeoff distance is affected by degree of slope as follows: For each degree of downhill slope, subtract 5 percent from your takeoff distance. For each degree of upslope, add 10 percent to your takeoff distance. These figures imply that an uphill takeoff may never be possible at some small, high runways.

By figuring the effect of a head wind or tailwind on takeoff distance from your operator's manual and then recalculating based on up- or downslope, you can come up with a theoretical preferred direction of takeoff. Having just given you a rule of thumb to compute takeoff distances in this instance, I will now advise you to exercise a *high* degree of caution in actually using that advice.

An uphill takeoff is most often a takeoff toward the head end of a valley. If the wind is blowing strongly down valley, forcing an uphill takeoff, you are almost certainly flying right into a region of very strong sink and a box canyon. If you must take off that way, the earlier you can get turned around, the better. It might be wiser to accept a downhill downwind takeoff in order to be aimed away from higher terrain and sink. It might be even more prudent to wait for a better day. Local conditions and local knowledge will strongly influence your decision. *Ask experienced local pilots who fly your type of aircraft.*

In the case of a very extreme slope, the basic rule of thumb will break down. On a steep enough upslope, your small aircraft may never be able to accelerate to takeoff speed. Once airborne, it might be unable to climb out.

On a steep downhill takeoff, your aircraft can very easily become airborne at lower than its normal rotation speed. Great! But don't press your luck. Once airborne, you must remember to leave the nose down and fly in ground effect until normal climb speed is reached.

Soft field (grass, dirt, snow)

Depending on aircraft type, most operator's manuals will advise adding 5 to 10 percent to calculate the takeoff distance for soft surfaces. While each individual make and model of aircraft has its own soft field takeoff procedure, what they all have in common is a takeoff roll with more flaps than a standard takeoff and the nose held high. The aircraft will become airborne at lower than its stall speed and will then be flown in ground effect until the normal climb speed is reached. At altitude you have to deal with reduced available power and need increased ground speed to achieve the necessary airspeed to fly. Because the power is so reduced at altitude, and because the takeoff is so prolonged anyway, just plowing through a soft surface to achieve the proper speed can be very difficult. Getting off the ground could take a great deal more than just a 10 percent longer distance. Indeed, on a very hot day at very high altitude, in soft dirt or sand, you could find yourself in a situation where, unless a good bounce got you airborne, you would *never* achieve takeoff speed.

This problem, by the way, is not limited to very high airports in the mountains. In soft sand on 100° days in the desert, with a heavy load picking up river runners in Utah, I have on occasion had to make multiple attempts to get airborne, usually relying on a good bump in the runway. A rule of thumb is that if you have not achieved two-thirds of takeoff speed at midfield, abort the takeoff and try again. Experienced backcountry pilots have often placed some kind of marker at midfield for this purpose.

Pilot skill becomes a much more significant factor in making the soft field takeoff procedure work at high altitude. Too nose-high an attitude on the takeoff roll increases aerodynamic drag significantly. Too flat an attitude increases the drag of the wheels in the soft surface. Very precise attitude control is needed to balance these two forces. For most tricycle gear aircraft, the proper attitude is to hold the nose high enough that the nosewheel is just above the surface of the ground by a couple of inches. It can be very helpful for the pilot to sit at the controls (with the engine off!) and have a friend push down on the tail until the nosewheel just lifts off. The pilot then stores the mental picture of this attitude for the actual takeoff.

Go/no go?

Most mountain airports are surrounded by high terrain, often high terrain in all directions. Before you decide to depart, can you outclimb the terrain? This is partly a question of climb angle, but it is equally a question of lift and sink. After takeoff, do you anticipate that you would be flying into some descending air? If that happened, could you climb through it? If you couldn't, where would you turn? These questions have to have answers before you depart.

Rethink the typical flight plan you learned in flight school. There you probably learned to take off, turn toward your destination, and start climbing. In Telluride or Aspen, to name two of hundreds of examples, on a flight to the east, you would likely depart west down the valley and not turn east for at least 5 mi, if not more, until you had climbed higher than nearby terrain.

Given the greatly increased takeoff distances at altitude, and the fact that at least some of your takeoff distance calculations may include some degree of fudge factor, it is possible that you may not in fact have enough room to take off when you thought you did. The decision to take off should be based not only on runway length, but also on the presence of nearby high terrain and obstructions you will have to clear.

One relatively easy calculation to do is your rate of climb. In fact, the aircraft manufacturer has provided you with rate of climb tables in the operator's manual. It will be helpful to convert climb in feet per minute to climb in feet per mile to see if you can expect to clear obstructions. From the climb tables, if your rate of climb at sea level is 1000 ft/min and your best climb speed is 80 knots, then in 1 min you will have flown approximately 1.3 mi from the airport and you will be 1000 ft higher than the runway. Is there an obstacle that high that close? Usually in the mountains there is, although surely you are planning to go around rather than over it. Why?

Go back to the operator's manual for the same aircraft at the 8000-ft-plus elevation of a typical mountain airport: Best climb is still an indicated 80, but this is a true speed of 96. In 1 min you will be 1.6 mi from the airport. Your rate of climb is now only 500 ft/min (if not substantially less), so you will only be 500 ft high.

Or play with the numbers this way: The obstruction is going to arrive 10 s sooner, and it is going to seem twice as high. The thing to be aware of here is that obstructions you would never even notice at sea level are going to be way too high to climb over. In fact, your rate of climb is so poor that very minor downdrafts will all but eliminate it. A 250 ft/min downdraft is not a big deal compared to your 1000 ft/min sea-level climb. It is one-half of your 500 ft/min high-altitude climb, and for many small aircraft, the calculated climb rate might only be 250 ft/min. You would effectively not be climbing at all. Your plan for takeoff has to include leaving room to maneuver around all sorts of seemingly insignificant obstructions, and even leaving room to completely turn back.

But assuming you have calculated that you can take off and you have begun your takeoff roll, what next? What if it seems to be taking more distance than you anticipated? Again, remember this rule of thumb: You should abort the takeoff if you have not reached two-thirds of takeoff speed at one-half of the runway distance. Before you take off, note where the center of the runway is, and make a mental mark of that point. If your takeoff speed is 60 knots and you haven't reached 40 knots by midfield, pull off the power and taxi back for another try (or wait until it gets cooler the next morning, or off-load some baggage).

Take off twice

This does not mean to take off only to bounce back onto the runway, but rather to take off two times with a reduced load. Although Glenwood Springs at 5916 ft msl is not an exceptionally high airport, its 3300-ft-long runway can be very short for that altitude on a hot day—too short for quite a few small aircraft to safely take off with full tanks and a full load. The solution? Take off twice!

You could elect to take off with all your passengers, but with minimal fuel. You would then land at Rifle, Colorado, about 20 min west,

500 ft lower, with a runway twice as long and fill the tanks there. Alternately, you could—and many people do—ferry two passengers over to Rifle, fill the tanks at Glenwood and pick up the third passenger, and then pick up the first two passengers as you go by Rifle headed west.

This strategy is almost always a necessity at many small private dirt strips in the mountains. Another idea is to check which airport has the cheapest fuel and figure out a way to ferry the people in a way that has you leaving there with full tanks!

Departures

Most mountain airports are located in the bottoms of somewhat narrow valleys. Your departure route is necessarily dictated by terrain clearance. A standard departure procedure or standard traffic pattern may not be wise or even possible. In calm conditions, it is pretty straightforward to figure out a climb that will miss everything. Usually this simply means a departure down the valley toward lower terrain. But you should also consider the unique effects of local winds with regard to lift and sink, and turbulence. The very best advice is to ask a local pilot about local conditions. A local pilot with considerable regular experience at any specific airport will have valuable local knowledge that a visitor can only guess at.

Some examples of specific local conditions at some mountain airports will give you an example of the kind of things to look for:

Glenwood Springs, Colorado, has a north-south runway at the bottom of a deep, narrow canyon. With strong prevailing west winds, the wind will spill down first one side canyon just south of the runway, then down another just north of the runway. Wind direction at the airport will be constantly switching. You could be hit with a very strong downdraft right after takeoff in either direction. The safest procedure in those conditions is to fly over against the east side of the canyon as quickly as possible. There, the blasts out of the side canyons on the west side will eventually be lifted up the canyon wall on the east side.

The *Telluride, Colorado,* runway sits on top of a cliff running the length of the entire south side, with lower cliffs off both ends. In a southwest wind, you could be airborne in a normal distance, only to be taken back down to the pavement after midfield by rotor off the cliffs. You can only hold your normal climb attitude and hope for the best. Since the center of the runway at Telluride is 78 ft lower than the departure end, either you will be taking off going uphill, or if you get airborne at midfield, your climb in a low-powered plane on a hot day will be about the same angle as the uphill gradient of the runway. How much of a fudge factor do you want to apply to this takeoff calculation? Maybe you want to allow quite a bit of extra room. Once past the end of the runway, the deep, narrow canyon below you is a likely collector of sink and turbulence, while the steep slope up to the north might be producing ridge lift.

Departing to the east at *Telluride,* the south at *Aspen, Colorado,* or the south at *South Lake Tahoe, California,* has you departing over rising terrain up into a box canyon. This is bad enough in itself, but is actually far worse than it appears because the same conditions that produced the down canyon wind that required a takeoff in that direction are also producing strong sink all over the valley ahead of you. Local pilots might advise you in those conditions how to execute a turn back out of the canyon as soon as possible at lower than normally desirable altitude rather than continue into ever stronger sink—or they might just advise you to wait until the next morning when the wind might shift.

The Jackson Hole, Wyoming, Airport is somewhere in this cloud of low fog from the Snake River. On this beautiful sunny day, the last three arriving airliners have all had to miss the approach and divert to another destination. Shelby Evans photo.

Kremling, Colorado, is in an open valley, but just downwind from the towering Gore Range Mountains. Local pilots could warn you that with strong west winds aloft, you could be climbing out into an absolutely horrendous rotor, and they might be able to advise you to choose an alternate route farther east.

Even the metropolitan Denver area on the wide expanses of the Great Plains is very close to the mountain front. Local pilots and Flight Service will warn you of sometimes very strong turbulence and downslope winds out of the mountains and will advise an initial climb out to the east over the plains.

Traffic

Most mountain airports are in narrow valleys. The normal approach to the airport is up the valley. The normal departure is down the valley. Nearby high terrain often makes a standard traffic pattern impractical, if not actually impossible. Thus, *opposite-direction traffic on the same runway is the norm, not the exception at a mountain airport!* This is no problem if you are expecting it, but keep your eyes open. The mix of large jet aircraft and small planes and even gliders at upscale resorts with no control towers invites problems. While the rule is to see and avoid, and while radios are not required at these airports, the practicality of the situation is that some of the bigger jets really have their hands full just getting to the end of the runway. Since you can maneuver and they can't, allow them extra room. Most large jets have TCAS, which is radar that picks up your transponder signal and lets them detect you from farther away; but don't count on what you hope someone else can do to cover your safety.

It is incorrect to assume that because it serves a small community, a mountain airport is not busy! Resorts generate considerable traffic. Aspen, Colorado, is the second busiest commercial airport in the state after Denver International, but busier than Colorado Springs. In May or November at Telluride, there have been occasions when my two or three takeoffs a day have been one-third of that day's flights, with two of the other flights being airliners. But at the height of the tourist season, Telluride with a population of only 2000 people can see more total traffic *and more jet traffic* than any western Colorado airport except Eagle or Aspen. Once at the very end of Christmas vacation when I was leaving Aspen, the ground controller told me I was number 22 to taxi. Thinking I had misheard him, I replied, "Number twenty-two to take off." Negative replied the ground controller; you

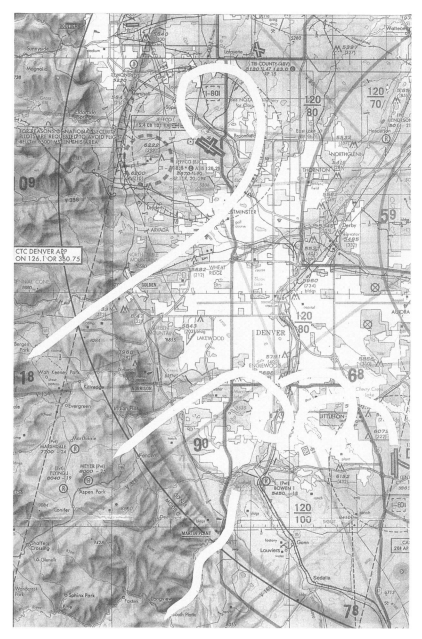

Leaving the Denver Basin: Strong westerly winds translate to strong sinking air over the foothills west of Denver—too strong for a small plane to outclimb. How to fly up over the mountains (from top to bottom): A circle east of Jeffco Airport at the top of the map gains extra altitude before you fly into the likely area of sink. A few circles of climbing near Centennial Airport, middle of the map, do the same thing. Finally, you can always climb up the Platte River canyon rather than try to climb directly over the mountains.

are number 22 to *taxi*. There were already 27 aircraft on the taxiway waiting to take off, all large jets. There were at least that many large jets making instrument approaches, and no other planes could be allowed to depart while anyone was on approach. After 3 h I had still not been given clearance to taxi, and so I decided to tie down the aircraft and try the next day. This was all for a 30-min flight all the way down the valley to Glenwood Springs.

A word of caution about your radio: The frequencies used by aircraft radios are essentially receivable only by line of sight. That means you cannot transmit or receive around a mountain. You can be calling out your position every few miles, but if you are still around the corner coming up the valley, someone about to depart in the opposite direction on the same runway you are landing on won't be able to hear you until you have the runway in sight. You won't be able to hear them either. Keep your eyes open.

En route

Route planning

In flat country your route planning takes two forms; either you go direct to your destination, or you go VOR to VOR. Neither of these is likely to work out for you in the mountains in a small plane. Even on a perfect day, a direct route over the mountains is seldom optimal. Altitude considerations may make it impossible. Even on a perfect day, you might not be able to out climb terrain on a direct route. The large number of crashes at Monarch and Independence Passes in Colorado testifies to this.

Without supplemental oxygen, you can only pop up over 12,500 ft for 30 min and can't legally go up over 14,000 ft at all. A direct route over Colorado, a state with 52 peaks higher than 14,000 ft, isn't going to work. Chances are that the optimal cruise altitude for your nonturbocharged aircraft is lower than 10,000 ft, which would make it most reasonable to fly down in the valleys. You might be flying so slowly during the long climb up to clear, high mountains that it would be faster to just go around.

Finally, weather, storms, wind, and turbulence might make it prudent for you to plan a route that takes the long, long way around. Mountain obscuration, rain, convective buildup, orographic snowfall, and lightning all form over the high peaks while the wider valleys remain

By choosing the upwind side of this Idaho Sawtooth Mountains ridge, the pilot is flying in smooth, lifting air. The other side is rotor turbulence and sink. John Kounis photo.

flyable. Turbulence and sink are an ever-present concern on the lee sides of any mountain, pass, or ridge. All these weather troubles can be minimized by electing to fly very early in the morning.

With the typically very high winds aloft of winter, I would seldom fly a small, light, four-seat or smaller aircraft anywhere but low down in the valleys out of the wind. In a larger turbocharged aircraft, I would climb directly to an altitude at least 2000 ft above the ridges. For whatever reasons, in a small aircraft, your route of flight in the mountains is typically very indirect. You tend to fly up and down major valleys and pop up over the passes in between.

The Colorado Pilots Association teaches mountain flight over a series of standardized routes. It is significant that *none* of these routes is described as "X airport to Y airport." All are described as "X airport to Y airport via Z Pass."

Navigation

As you learned when you became a pilot, there are four basic methods of navigation: pilotage, i.e., looking for landmarks; dead reckoning, i.e., those calculations you did for wind correction angle with the circular calculator; radio navigation, as with VORs; and satellite navigation with a Global Positioning System (GPS). Ideally, you will employ all these methods simultaneously. Be aware, though, that not all these methods will work particularly well in the mountains.

Pilotage.   Probably not much emphasis was placed on pilotage during your training. How you do it is self-evident anyway—you look out the windshield and identify landmarks. There are places where pilotage doesn't work well. Over western Kansas, for example, all the towns are about the same size. Each one has a grain elevator, and all are painted white. If you drop down low enough to read the signs on the water towers, you learn that one-half of the towns in Kansas are named "seniors 99" and the other half are named "Jesus saves."

In the mountains, though, major river valleys are often easy to distinguish. Only some valleys have highways. At least if you live in the area, certain mountain peaks are very distinct. There are many very unambiguous landmarks, and your route of flight likely follows some major valleys. Indeed, the lower the performance of your aircraft, the more likely it is your route will consist entirely of following major valleys, which is coincidentally where the only major highways are located. Weather and high terrain will force this option on you.

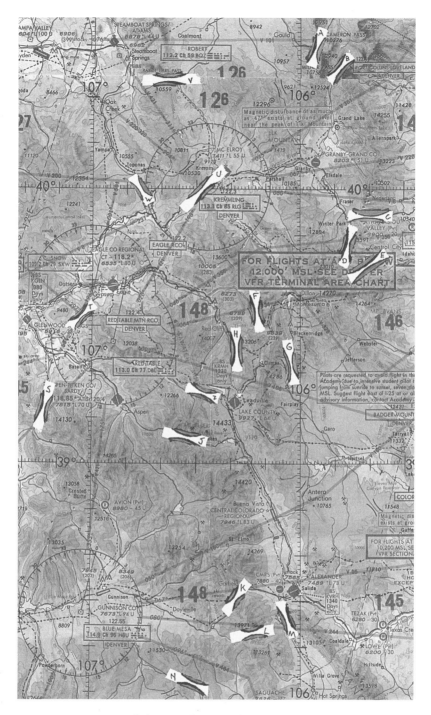

Mountain passes: All the small aircraft traffic crossing the 14,000-ft Colorado Rocky Mountains will necessarily choose this very small number of lower mountain passes.

161

Particularly for mountain residents who are familiar with their local area, I recommend pilotage as the primary means of navigation. If you are not really intimate with the route, plan beforehand. Mentally (or with a pen) mark significant features on the map which will help you make a positive identification of the valley you are looking for.

Baker and Lamb, in their landmark 1989 study of aircraft accidents in the Aspen, Colorado, area, noted that several pilots approaching Aspen from the east simply misidentified the valley leading toward Independence Pass and flew up the wrong valley. Do you want to avoid that particular mistake? Check your sectional chart: The valley leading up to the pass has a highway and two very large lakes at the foot of the pass. The valleys to the north and south don't.

Dead reckoning.   Dead reckoning is a calculation based on your traveling at a known airspeed on a known heading. A correction is made for a known wind direction and known wind velocity to give you a calculated ground track at a calculated ground speed. The *dead* in dead reckoning has nothing to do with death (you hope); rather, it has to do with deduction. Think of *dead reckoning* as being short for *deduced* reckoning. This procedure works reasonably well, if imprecisely, over the ocean. It can work reasonably well over the vast expanses of the desert where VORs are few and so far apart that there are gaps in reception. Charles Lindberg crossed the Atlantic using only dead reckoning, dropping down to near wave top height every so often to get a new wind estimate. The fact that he did find Paris testifies to this method's possible accuracy. You probably spent a lot of time doing these calculations with your circular calculator in preparation for your private pilot written test.

Unfortunately, this method of navigation can be nearly worthless for a small, low-flying, slow aircraft on a long trip in the mountains.

Let's take it factor by factor. You are almost certainly not flying at your assumed airspeed. As you fly through sink, you have to climb, so you end up flying more slowly than predicted cruise. When you fly in lift, you put the nose down a little and go faster. In turbulence you deliberately slow down. On a long descent if the air is smooth enough, you go far faster than anticipated. In turbulence you find that you are constantly adjusting the power setting. Your speed in flying a small plane in the mountains is never constant. Soon, you have no idea how close to your predicted speed you actually flew. While you planned to fly a specific heading, you soon find you are maneuvering around things you expected to climb over. You are making small course changes to deal with little clouds and turbulence. You are no longer flying anything like a constant heading. Wind direction is not constant and not close to what is predicted. Wind is deflected by ridges; it is channeled up and down valleys rather than across them. It deflects around peaks and venturis through passes. Nor is wind speed constant. The wind in sheltered valley bottoms is calm. Over the tops of sharp ridges and through narrow passes, it can be double the predicted speed. It can be blowing east up one valley at 5 mi/h, south down another at 10 mi/h.

All four factors you used to make your dead reckoning calculation turn out to be very different from those predicted. The error introduced by these combined inaccuracies is more than enough to have you in the wrong valley altogether, if not three valleys off.

Baker and Lamb note that a contributing factor in a couple of crashes east of Aspen was that by using dead reckoning calculations coming up the Arkansas valley, pilots who thought they were going to turn west and cross Independence Pass were in fact one or two major valleys farther north or south.

Nevertheless, dead reckoning has a very useful, very valid application for flight in the mountains. That application is to simply use it for shorter individual legs of a flight. On a flight up the Arkansas River Valley to Aspen, Colorado, which began as a flight up the Upper Rio Grand River Valley from Albuquerque, New Mexico, your dead reckoning position 2 h into the flight when it came time to turn west and cross Independence Pass would be hopelessly off. On the other hand, if you have correctly identified the very large town of Salida, Colorado, your dead reckoning calculation for flight time up the valley for the 20 min could be a very valuable aid in choosing the correct pass over the Continental Divide.

VOR navigation. This form of navigation is based on receiving a radio signal from a ground station. No doubt you became very proficient at this while preparing for your first check ride, and you became a master of the art if you got an instrument rating. Flying over flat country, you can track VOR radials from stations over 100 mi away. Unfortunately, as with communications, these systems use radio frequencies that work only by line of sight. If you are flying below ridge top height, you cannot receive the signal from a station only 5 mi away over on the next ridge. VOR stations are expensive to build and maintain, so there are not very many of them in the mountains. Over much of your route, you will be flying too low to pick up even very nearby VOR signals. The minimum en route altitudes for most instrument routes over the Rocky Mountains are over 17,000 ft. Below those altitudes you often lose the signal. Because of the costs of maintaining these stations and because of the widespread use of GPS, the VOR system was originally scheduled to be deactivated in 1999 and probably will be deactivated gradually over the next decades.

The Victor Airway system of routes connecting VORs exists with little regard to the terrain below the route. The fact that a Victor Airway crosses a mountain range does not in any way imply that your particular airplane should be flying that route at low altitude. It only means that the mountains lie under a straight line between two VORs. Baker and Lamb interviewed pilots who crashed on Monarch Pass, Colorado, who had assumed that because there was a Victor Airway crossing it, the pass should be safe. VOR navigation will work for some portions of some routes, so it is only prudent to have the nearest station dialed in; but it definitely will not work over all portions of all routes.

Again, the entire VOR system was to be decommissioned in 1999. That date has been pushed back and will no doubt be pushed back again, but someday the system will be decommissioned.

LORAN. LORAN signals are broadcast on a radio signal of 100 kHz. This very long frequency will reach over peaks down into valleys. LORAN was invented for marine navigation and adopted by aircraft. It was a great system near the coast where there were lots of stations. Coverage in the mountains was poor, but improved with the commissioning of the midcontinent chain of stations. Many aircraft still have these receivers, and the system works reasonably well in the mountains with only a few areas of no reception.

LORAN, too, has had its decommissioning date put back a few times, but will eventually be gone. Very few aircraft have their LORAN receivers anymore, particularly since avionics manufacturers began offering slide-in GPS replacement units.

GPS. The Global Positioning System has been around for a while, but unless you learned to fly very recently, you probably didn't receive much instruction in it. Your check ride dealt only with VORs, and GPS was regarded as cheating. You probably got a GPS only after you got a license. Use it! GPS works great in the mountains because it derives its information from satellites overhead, so it still receives signals down in the valleys. GPS may not be perfect. The signal can degrade with sunspots or solar flares. The military can degrade the signal in an emergency. There is little standardization, so one GPS unit does not work like another. But all in all the system is still amazingly accurate and easy to use.

Combined with pilotage, the most basic GPS information about direction and distance to nearby airports makes mountain navigation very accurate and very easy to master. But you *do* have to keep using pilotage because your GPS will happily direct you right through a mountain to your destination.

Most pilots fall into the habit of just entering their destination in the GPS and never getting out the chart. Particularly in the mountains where the GPS database has no idea about the local terrain, using GPS to locate your position on the chart is far safer and scarcely more difficult than using just GPS alone.

Various GPS displays feature moving maps. While most of these are limited to a display of just a few waypoints and routes that correlate to instrument en route charts and approach plates, there are systems available that display a full color rendition of current sectional charts. Perhaps sooner than we think we will have synthetic vision which is a computer-generated forward-looking graphic picture based on sectional chart databases and GPS positioning.

Aircraft designer Burt Rutan has said that we are the last generation of pilots who will learn instrument flying. He predicts that in a decade or so, all new aircraft will have a GPS-based heads-up display that will depict the terrain ahead. All your flights will just be VFR looking out the windshield. Sometimes you will see the ground, other times you will see a computer image of it.

Use all the navigation methods available together! Each navigation method has its own shortcomings. Three student pilots are about to provide us with examples to show us why using more than one method of navigation at once is the best practice.

Pilotage only. A student pilot from Glenwood Springs in 1995, the year before I went to work there, was fairly familiar with his landmarks for western Colorado and still not very confident about other navigation methods. Since he had no problems with pilotage, he elected to use it exclusively. His flight from Glenwood to Grand Junction, Colorado, following the valley of the Colorado River and an interstate highway was uneventful. His return would have been equally simple but for one thing. The control tower at Grand Junction instructed him to depart from runway 22, but his plans had been for runway 29. Once airborne, he made about what he considered to be a turn en route and began looking for landmarks. Sure enough, there was a river and a highway, which he duly followed. After a time he saw an airport, which he identified (incorrectly) as Rifle, Colorado. Now the canyon should narrow, and it did.…But where was Glenwood Springs? Continuing onward, he crossed a low divide. Still no Glenwood. And then he crossed a range of mountains. This definitely could not be right. What had happened to Glenwood Springs? Realizing fuel might run low, he wisely decided to land at the next airport he could find. In the event, this proved to be Alamosa, Colorado, 200 mi southeast of Grand Junction instead of Glenwood, 70 mi northeast. The student pilot had simply followed the wrong valley, the wrong river, and the wrong highway, and when things didn't look right, he kept going. The decision to stop and ask directions was a prudent one.

Dead reckoning only.   In 1997, the first year I worked in Telluride, a student pilot from Montrose, Colorado, headed out on her first long solo cross-country to Farmington, New Mexico. This particular student had a mild block about "technological" things such as radio navigation, but was a whiz with simple mathematics. Dead reckoning calculations flew from her fingertips like musical notes from Little Richard at the piano, so she elected to use that method. Finding Farmington was easy, but the return flight was endless. In fact, the time was up but still no airport. She wisely decided to keep the only road she could see within gliding distance. This was a good idea, because when the engine eventually ran out of fuel, she was able to make a very smooth landing on the road and hitchhike to the nearest phone. What happened? While she was away from the plane at Farmington, the aircraft was repositioned on the ramp. The heading indicator was now close to 180° off, but she didn't notice and flew the course she had previously calculated, paying so much attention to heading that she didn't notice landmarks out the window. She ended up hundreds of miles the wrong way out on the Navajo reservation.

VOR navigation only.   VOR navigation isn't absolutely foolproof, but this particular student had certainly mastered it well enough. She tracked outbound on a radial toward Moab, Utah, in 1996 with no difficulty. Problems began once she tried to acquire the Moab VOR. Now there is indeed a Moab VOR on the sectional chart and in the airport facilities directory. What any local pilot can attest to, though, is that for some reason the Moab VOR seems to be out of service at least 99 percent of the time, as indeed it was on this particular day. Had you been thinking pilotage, you would now search along the highway for the airport. Were you this particular student and totally committed to the VOR, you would be lost. Obviously it would have been a good idea to call Flight Service before the flight to find out if the VOR was in service. Have you ever departed on a flight without doing that first? Gee, none of us has either. Right!

Lost?

This is as good a place as any to review what you should do when you are lost, because this particular student did exactly the right things and thereby solved her problems.

1. *Confess:* She admitted she was lost and started to take action.

2. *Conserve:* If you don't know where you are, then you should be in no hurry to press on. She throttled back to save fuel.

3. *Climb:* Climbing up even just 1000 ft makes it much easier to identify landmarks, and she did.

4. *Communicate:* She was lost in that she could not find the airport, but not so lost she had absolutely no idea where she was. Accordingly she made a call for help on the Moab airport unicom frequency. (You could try Flight Watch on 122.0 if that got no response. Flight Watch would give you a Center instrument controller's frequency, and the controller would get your transponder on radar and vector you toward an airport.) She said something to the effect of "I'm lost, circling just west of the highway somewhere near the Canyonlands Airport climbing through (some altitude) and all I can see is rocks." Bonnie Lundgren, then the Moab Airport manager, came on the radio and instructed her to circle and climb a little higher. The pilot did and seconds later Bonnie radioed, "I have you in sight about 3 mi south of the airport." The adventure came to a happy conclusion.

By the way, all three students in these examples have their licenses now, navigate very efficiently using a variety of techniques, and would be very safe to fly with. All their instructors now stress using multiple navigation methods.

En route flight techniques

If you planned well and the weather is good, you can certainly get away with not using, or even knowing, these techniques. You know and practice them in good conditions because you cannot accurately predict mountain weather, but you still plan to avoid the situations where these techniques are required. Safe en route mountain flight is mostly dependent on the early identification of hazards rather than possessing any special technical flying ability.

You have already determined that your route of flight will follow major valleys and pop up over mountain passes between them. You also remember that for a variety of reasons you are almost assured of encountering sink, and that in the wrong spot you could find sink rates that actually exceed your ability to climb. If the weather is anything but severe clear, then you are aware of the possibility of rapidly developing clouds obscuring some of the mountains. You are definitely planning to go, but every time you fly, you are planning to be cautious.

Some hazards are obvious enough that everyone is aware of them and few accidents result. For example, worldwide, inadvertent flight into instrument weather conditions year after year is the biggest cause of general aviation accidents. The danger of this situation when you are surrounded by jagged mountains is obvious. About the time the mountains themselves begin to be obscured, most people wisely decide to find a nearby airport and land before things get worse down in the valleys as well. Despite the extremely rapid rate at which weather can deteriorate in the mountains, this hazard is so obvious in the mountains that few pilots risk it and relatively fewer accidents result.

For one reason or another, it is the inability to outclimb terrain that is a factor in most small aircraft accidents in the mountains. If the air never moved, this might not be the case. It is my hope that if more people are aware of the nature of the problem, more of them will take some relatively easy steps to address it.

Dale Wood is a pilot based in Telluride, Colorado, and the chief of police for nearby Mountain Village. Some years back while serving as police chief in Grandby, Colorado, he was called upon to investigate the wreckage of a small aircraft found by backpackers. The aircraft was a Cessna L-19, a military surplus artillery spotter which saw extensive service in southeast Asia. It is an agile, little two-seater with a very big engine and a climb rate better than double that of the Cessna 172 whose wings it used. This should be a far safer than average small plane in the mountains.

The Cessna l-19 Bird Dog crash investigated by Dale Wood. A video camera in the plane recorded the crash. The wreckage was not discovered for 2 years, in part because the fuselage is upside down, lying on top of the ELT antenna. Dale Wood photo.

The wreckage had gone undiscovered for 2 years because the plane went in vertically in tall trees. No emergency locator transmission (ELT) signal was transmitted because of debris on top of the antenna. The plane had gone missing in severe clear weather on a calm day. What had happened? From the wreckage Dale recovered a smashed video camera with a tape still inside. It is chilling to watch.

The tape depicts most small aircraft mountain accidents. The video camera is mounted looking forward. Most of the sound is engine noise, but the two people's conversation is very clear. They are flying east toward the higher mountains of the Continental Divide. The scenery is spectacular, and that is the subject of all the conversation. They fly over some pristine lakes and remark on them. The snow-covered mountains of the divide are far distant. Only if you already know what is coming do you really notice that the ground is gradually creeping upward.

The crash itself is over in less than a second. Suddenly the ground in front of them rises slightly more steeply. The pilot shouts, "Oh, my God." He tries to turn back but is already too close to the treetops.

You hear the stall warning, there is a stall, one wing snaps down into a 90° bank, and they hit the ground before the spin can even develop.

With no warning that they were aware of, they were very suddenly too low with no way to turn around. About two-thirds of all small airplane accidents in the mountains include some portion of this scenario. In all probability, if you are reading this book, you are flying an aircraft with one-half their climb rate and will someday find yourself in weather worse than they encountered. *But you are not going to crash, or even scare your passengers.*

That is because you are going to be willing to turn around, you are going to know when to do so. You are going to be prepared to make the turn early even when you don't believe it will be necessary, and you will be familiar with turning techniques.

This accident nails pilots visiting the mountains every year. It seldom catches experienced mountain pilots because they automatically prepare for it every time they fly. That is not to say experienced pilots are never caught in this trap—they are just more cautious about it and get caught less often.

Narrow valleys

Your route of flight will frequently, if not invariably, find you flying up a valley or canyon, intending to cross the pass at the head end of the valley. At first it will seem there will be plenty of room. But the day will come, perhaps as soon as your next flight, when you will fly into a little descending air and won't be able to clear the pass. You will have to turn around. This is how to do it.

Climbing a narrow valley

As soon as you begin to ask yourself whether you are going to out-climb the pass at the end of the valley, it is *already* time to turn around and gain some more altitude before the next attempt. At the very core of mountain flying skills is the ability to force yourself not to push your luck and not to try things you think you can only probably get away with.

The first question to ask yourself as you start the climb up any valley is, How will I turn back?

Flying a narrow valley. Notice that the pilot has elected to fly one side of the valley even though his turn radius does not really require him to. Good habits become very automatic. John Kounis photo.

Even though he is about 1000 ft higher than the pass he is about to cross, this pilot is still beginning a turn that will result in his crossing at a 45° angle to the pass. John Kounis photo.

You have to choose a route of flight that gives you the maximum amount of room possible to turn around. That means climbing glued right up against one wall of the valley. You can fly very close to terrain beside you. If you start to lose altitude, you just turn away from it. (You cannot fly safely at all close to terrain below you because if you lose altitude, there is nowhere to go.) The very safest place to be is usually extremely close to the side of the valley because then you have the most room possible to turn around. In practicing this procedure with flatland-based pilots, when I say get close to the side of the valley, they generally elect to fly a distance of just over a mile away. In a normal (whatever that means) valley, a distance of under 300 yd is optimal. In a very narrow valley, less than 100 yd might be required. In those situations where search and rescue flights to save someone's life have you looking into places you would not otherwise go, *only a wingspan or two off the cliff face is not out of the question if winds are very light and the canyon is extremely narrow. You need to leave as much room as possible to turn around!*

One side of the valley will definitely be better than the other because one side of the valley will have lifting air; the other may have sink and possibly turbulence. If the prevailing wind is across the valley,

This Bonanza pilot is also flying close to the cliffs up one side of the Crystal River Valley leading to McClure Pass, Colorado, leaving ample room to turn back—even though the photographer wanted him closer to the camera plane for the shot! John Kounis photo.

the upwind side will have air spilling down into the valley, and the downwind side will have smoother air compressing together and rising back up. Turning around from the downwind side of the valley is a turn into the wind, which will use up less room than a downwind turn. If one side of the valley is getting a lot of direct sun and the other side is in deep shade, the sunny side will have convective lift.

For whatever reason, fly the smoother, lifting side. (*Note:* This rule applies except in very special circumstances, discussed much later in Chap. 8)

Did you guess wrong and find yourself flying up the turbulent, sinking side of the valley? If there is room to do so, quickly cross over and try the other side. *Note, however, that the "wrong" side of the valley is still much better than flying up the middle of the valley* because it still leaves you more room to make the turn. In fact, the wrong side of the valley will alert you to the need to turn back sooner.

Climb speed

Variable V_y is your aircraft's best rate of climb, V_x is its best angle of climb, which is a slightly slower speed. Most texts will tell you to

Heading down *the valley east of Independence Pass, Colorado, this flight instructor is already headed toward lower terrain and so does not need to keep to the side of the valley anymore.* John Kounis photo.

climb at V_x to clear an obstacle. I am going to tell you not to, at least in a small, low-powered aircraft in the mountains. In climbing in the mountains, your concern is not so much the climb angle itself as it is your ability to maintain that angle in unexpected turbulence or sudden encounters with sink. Flying slower than V_x means you are not achieving your climb angle at all—you are just flying around with your nose very high in the air near a stall. The normal variations in airspeed resulting from very light chop, and flying in and out of very small variations in lift and sink produce enough momentary changes in airspeed to completely overcome your attempts to maintain the best climb *angle*. Flying typically 10 knots faster at your best climb *rate* protects you from a turbulence-induced stall and means your actual airspeed is now varying between best climb angle and a little faster than best climb rate. You will be achieving something actually closer to best climb angle than you could when trying to fly right at your best climb angle speed.

Climb speeds change with load and altitude. Most people tend to think of the best climb rate as one specific speed. It is not. Read the fine print in the operator's manual. The best rate of climb V_y is at a slower speed with a lighter load. Also V_y is a slower speed at higher altitudes. Let's look once again at the ubiquitous Cessna 172: Best rate of climb at sea level at full gross weight is 80 mi/h. At 5000 ft at full gross weight, it is 78 mi/h. At 10,000 ft flown with two people on board and no luggage, V_y is 74 mi/h. Flown solo at 15,000 ft, best rate of climb is achieved at only 70 mi/h.

Clear the pass at the head of the valley?

When I teach this procedure to students, their first reaction to this question is to ask the altitude of the pass. I won't tell them. The reason is twofold. First, pressure varies greatly from airport to airport over the mountains. The altimeter setting you are using could be 200 ft off. Second, and more importantly, knowing your altitude in relation to the pass is less important than seeing whether your actual climb or descent path is going to clear it. When you look past the pass, are things in the background rising or sinking out of view beyond the pass? If more and more things are coming into view, then you are going to clear—*provided your flight path does not change.*

The last phrase is emphasized because your rate of climb almost certainly will change as you fly through lifting or sinking air. You are flying glued up against one wall of the valley because you are anticipating suddenly being flushed down the drain by a very strong unexpected downdraft, and you are already setting up to escape that eventuality. As soon as you begin to question whether you will be able to clear the pass, you should already be turning back. Make a big, wide climbing 180° turn, and gain some altitude before flying back up the valley for another try.

The overriding thought in your mind as you climb up a canyon is that you will *not* be able to clear the pass. It should come as a pleasant surprise if you actually do find yourself high enough (even if you do encounter this nice surprise over 99 percent of the time because of your excellent planning!) and no catastrophe when you don't because you are already prepared to turn back.

Canyon turns

Every mountain pilot seems to believe he or she has the last word on canyon turns, and that this skill somehow produces some sort of mountain invulnerability. Actually the closest thing to invulnerability is the judgment required to know when to turn early so that you do not have to employ a dramatic canyon turn.

The time will come when you are flying up a canyon and will have to turn back. It could be because of a downdraft; it could be because of bad visibility. It might just be the result of bad planning. But it will happen repeatedly. It has happened to every other pilot flying a small plane in the mountains so often that no one keeps track, and if you know what to do about it, the result is a complete nonevent. Here is what you do:

The smartest thing to do is to turn around early before the situation gets desperate. Then all you have to do is to make a big, wide turn at moderate bank angles. But for whatever reason, say you have let it go too long and now you are flying up a very narrow canyon and the need to turn around and get out of there is becoming critical. Maybe it wasn't your fault. Maybe it is the result of a very sudden and unexpected downdraft from one of those anomalous reverse-direction winds. Whatever. You personally are going to have to do something about it right now. Aren't you glad you practiced these things under less demanding circumstances? The turning techniques are discussed next more or less in order of increasing desperation.

Normal steep banked turn.   You demonstrated the ability to hold altitude in a 45° banked turn when you passed your private pilot check ride. Now you get to do one in real life. Steeper banked turns, say, 60° bank, will produce a smaller turning radius, but be careful. In a 60° bank while holding altitude, you are pulling 2g. If your stall speed was 57 knots flying straight ahead, it is 65 knots holding altitude in a 40° bank and 81 knots in a 60° bank. You can bank much more steeply than that without stalling if you don't pull back so hard on the yoke, but then you are not holding altitude and you are dropping deeper into the canyon where it gets narrower.

The slower you are going, the smaller the radius of the turn. If you are flying a faster, heavier, high-performance aircraft, it might be advantageous to quickly reduce power and add flaps. For a lower-performance, lighter aircraft, see the next method.

Climbing steep banked turn.   With practice, you should be able to develop the ability to fly a 30° to 45° banked turn at a constant airspeed. That speed should be your best rate of climb speed. This is just like the normal steep banked turn, except that because you are going up, the canyon is getting wider, and because you are going slower, the turn radius is smaller. For most pilots, this is the procedure I would recommend and the procedure you should practice. Turbulence and sink can make this difficult to do because your airspeed will vary greatly as you are hit with headwinds or tailwinds or up- or down-moving air. The exact bank angle to use for this procedure will vary depending on the type of aircraft. Lower-performance aircraft may not be able to climb at all at high altitude with too great a bank angle. Your situation will dictate whether climb or turn radius is a higher priority.

As I read about it in print, it seems that this should be a fairly easy skill for most pilots. With practice, it is. When I try to teach this skill to visiting pilots on their first mountain flights, the proximity of terrain and a degree of turbulence seem to mean that usually speed control is difficult, so they are going too slowly early in the turn and fall out the end nose-down with speed building up quickly. A very small amount of practice over flat country can go a long way toward building this skill.

Steep turns with flaps?   Either of the two methods of turning described above can be done with 10° or 20° of flap. Most aircraft will definitely turn with a much tighter radius with added flaps because you can make the turn at a lower airspeed. Whether you should do this depends very much on the aircraft type. A somewhat heavier plane with a larger engine can usually use this method to advantage.

The aircraft's momentum will cause it to balloon up from the extra lift when you add flaps, the extra drag will slow you down and thereby decrease the turn radius, and the more powerful engine will help you hold altitude or even climb despite the increased drag. With a very light, low-powered airplane, the drag of adding extra flaps may cause enough altitude loss to do more harm than good. Practice will help you determine the procedure best suited to your aircraft and altitude. I would typically add 10° or even 20° flaps in a 3800-lb gross weight Cessna 210, 206, maybe in a lighter Cessna 182, but probably not in a 2300-lb gross weight Cessna 172. This technique is counterproductive in a 1200-lb gross weight Cessna 150 or 152, in part because the flaps are of very narrow span. These aircraft have modified Fowler slotted flaps which greatly increase lift. In trying this same procedure in an old Beech Travelair twin with plain flaps, the only benefit seemed to be added drag slowing the aircraft—there seemed to be no improvement in lift.

The 40° flap turns.   The only reason I discuss this very frequently recommended method of canyon turns is to tell you emphatically

Photo courtesy Tim Martin.

not to do it. You can make a very convincing demonstration of this technique high over flat terrain not in a canyon. Someone else's student pilot demonstrated this procedure to a friend in a Cessna 172 and turned half the hair on his head gray. As with all bad examples in this book, the incident is disguised enough to hide the true identity of the participants, but the story is as exactly correct as I can make it.

The idea is that dumping in 40° of flaps throws on the brakes, balloons you up with extra lift, greatly reducing the turn radius because you are going so much more slowly. As an added benefit, your stall speed is lower, so you can fly the turn at an even slower speed or even steeper bank angle. On paper, this looks very good indeed, and when it is demonstrated not actually in a canyon, it appears to result in a 180° turn in a very small space.

With apologies to everyone advocating this procedure, including some very experienced Civil Air Patrol search and rescue pilots, and some devoted missionary workers, I will again emphatically say, *Don't do this!*

At high altitudes, it is easy to get a small, low-powered aircraft behind the power curve with full flaps. In this mode, you are going so slowly at such a high angle of attack that lifting the nose does not result in a climb. In fact, full power is required to just hold altitude. Lifting the nose actually *increases* your sink rate, so now you are going down with full power almost stalled. The only way to recover from this situation is to lower the nose. What happened when the student pilot demonstrated this goes as follows:

On the first attempt the stall warning was blaring, so they were about to stall in a steep turn and plunge very deep into the canyon. A Cessna 172 is very easy to get out of a spin, but not with full flaps. Even without a spin, stall recovery would have had them very near the rocks. My friend took the controls.

On the second attempt, wary of the near stall, the pilot kept the nose lower. This time they lost very considerable altitude, literally diving in order to maintain minimum airspeed, and found themselves too deep in the canyon for safety. My friend again took the controls and quickly raised the flaps to 20°.

On the final attempt, because they lost so much altitude on the previous try, the pilot held the nose higher and got behind the power curve. This time they were really headed down *and* near stalling before the turn started. Once again my friend took the controls, much sooner, and suggested that they stop doing this.

Later the student pilot confided that he had been practicing this solo all day the day before in a much narrower canyon and never did quite get it right. That was the precise moment when my friend's hair turned gray.

The pilot's instructor, by the way, is still teaching students how to do this and sending them out solo to practice. He believes that it ultimately produces safer pilots. Provided nothing happens to them during the learning process, he could be right, but perhaps they would learn this particular skill just as well with supervision.

Maybe, just maybe, at lower altitudes a skilled pilot in the right aircraft can make this work. Trying this procedure in a turbocharged, retractable-gear 182, we seemed to do very well, provided we started out with plenty of extra speed. When you factor in thin air, real turbulence, real sink, and a very narrow canyon, the odds of your being able to hold the precise speed and attitude necessary for success get very small.

Unlikely methods for canyon turns

While I am going to describe a few additional methods for making a canyon turn, I preface the next section by saying that these are methods which almost certainly won't work, and you should have long since turned around before it was time to try any of these. I am mentioning these methods because you hear them advocated so frequently if you hang out around the airport on weekend afternoons.

All these require a level of ability the average pilot can easily acquire, but probably does not have because she or he has not practiced these specific skills. All require entering the maneuver while carrying considerable energy in the form of either considerable speed, which is unlikely in this situation, or considerable horsepower, which is even less likely.

Chandelles

Chandelle means candle in French. During World War I, fighter pilots in France discovered that they could turn around very quickly in a very small space by using a specific type of climbing turn. They employed this technique to escape the guns of the Huns. They spoke of a climb *en chandelle*—a climb spiraling up a candlestick. (The Huns invented an even more effective countermove, the *Immellmann* turn, in which they performed half a loop and rolled right side up at the top.)

You have to demonstrate a chandelle on a commercial pilot check ride. Flying at or close to maneuvering speed, roll into a steep banked turn. Then gradually lift the nose into a climb. Maximum bank is reached at about a 90° change of heading. At that point, the bank angle decreases and the pitch attitude increases until at the end of a full 180° turn you are at zero bank and pitched nose up fairly close to a stall. If it is done properly, your turn radius over the ground can be very small, and you complete the maneuver at a much higher altitude than you began. Obviously if you are doing this in turbulence, you don't want to come too close to a stall at the end, even though that is one of the things which really tightens up the turn.

There are reasons to doubt that this procedure will always work. You have to start this maneuver in a shallow dive to build up speed. In a narrow canyon under duress, chances are that you are already flying very slowly at best climb speed and there is not time or space to dive for better speed needed to initiate the maneuver.

Half lazy eights

A lazy eight is one of the commercial pilot test maneuvers. After a shallow dive to get your airspeed up to maneuvering speed, you climb steeply and then begin banking. You increase the bank angle while decreasing pitch attitude. At the point where you have completed 90° of turn, you are at zero nose-up pitch and maximum bank angle. Bank is then gradually reduced, and nose-down pitch increases as you smoothly complete the rest of the turn. Then you repeat the whole, doing a turn in the other direction to get the entire eight pattern.

The idea here is that first you climb steeply and lose some speed; then because your speed is slow, the turn has a tight radius. Now you are flying very slowly, but regain speed by trading off altitude as you descend out of the turn.

Normally you would commence this out of a shallow dive to get up to maneuvering speed. If it is done that way, entry altitude equals exit altitude. This can also be executed from a steep climb at low speed, but you will inevitably be much lower coming out of the maneuver than entering it if you do it this way. That might or might not be a problem, depending on the nature of the terrain.

Wingovers

This is a lot of fun, but perhaps not always practical. Think of it as a much more aggressive half lazy eight. Climb much more steeply. At the highest point of the climb, you are halfway through the turn. Your bank angle is 90° (more if you really need to impress everyone), and your airspeed is very, very slow. It feels close to zero. Then you essentially fall into the second half of the turn.

As with the other, more extreme maneuvers in this section, high entry speed is required to do this right, and you are unlikely to be able to achieve it. Unlike some of the other maneuvers, you *can* accomplish this maneuver by starting at the lower speed of a steep climb, but your altitude loss will be dramatic. It certainly requires considerable skill and practice to do it well or consistently. There is a good chance that without such practice, you will not fall out of this maneuver in anything like the direction you intended. That would defeat the whole purpose of doing it in a narrow canyon.

As with half of a lazy eight, if you commence this maneuver out of a slow, steep climb, *exit altitude will be far lower than entry altitude.*

For both the half lazy eight and the wingover, a very powerful engine and a very lightweight aircraft will make things work much better because altitude loss will be less.

Hammerhead stall turns

This is really your last resort. I include it in this discussion only because so many pilots hanging out around the airport on a Sunday afternoon will advocate it. Performed by a practiced, expert pilot, this can result in a turn almost within the plane's wingspan with less than 50 ft of elevation loss from the altitude where the maneuver was initiated—but this can be a very demanding maneuver. The following is a description of how it is done, *not* advice on how to do it!

From a dive at or close to maneuvering speed, pull up aggressively and abruptly into a vertical or near-vertical climb.

Speed will bleed off quickly. Just a tiny fraction of a second before the moment your speed reaches zero, smoothly kick in full left rudder. As the aircraft pivots, add in full right aileron to maintain heading and keep from rolling or spinning. As you go past horizontal pitch, add in some forward elevator. You end up in a vertical dive exactly opposite your direction of flight when you pulled up.

Why *not* do this? Pulling out of the near-vertical dive at the end could stress your aircraft beyond its capacity and damage it. As you run out of energy in the vertical climb, engine torque will begin to overpower the control surfaces and start a roll to the left. If it is held vertical too long, a tail-first slide rather than a quick flip around into a dive will result.

In September 2000, I watched in horror as a very skilled air show pilot attempted this move in a 500-hp high-performance aerobatic airplane. To make a more dramatic show, he held it in the vertical climb just a fraction of a second too long, and just then his engine sputtered a little bit. Engine torque started a bit of a roll. He slid backward tail first. The aircraft snapped into a spin, and before he could complete the spin recovery he had smashed into the ground. By the time the ruptured fuel tanks exploded 10 s later, he was already dead.

This happened to a professional aerobatic pilot who practiced almost daily in an airplane designed and stressed for this kind of maneuver. How often do you practice this?

I don't mean to imply that this cannot be done—it certainly can, and with great effectiveness. I do mean to stress that this maneuver requires a very high degree of skill and a lot of practice. It requires an aircraft capable of withstanding the stress of the pullout from a vertical dive. The turbulence that requires you to resort to this maneuver is going to greatly complicate things.

This maneuver is also initiated at high speed, normally achieved by diving. Chances are good that by the time you need to do this, you are already climbing at best rate of climb and there is not time or room to dive for greater speed. If you do succeed, your altitude exiting the maneuver will be far lower than your entry altitude. Finally, if you get yourself into the situation where you have to resort to aerobatics to escape, you have put off turning back far too long.

All canyon turns

It should be obvious that the sooner you make the decision to turn around, the less demanding the procedure will be. Judgment rather than skill provides the highest degree of safety.

Flying up a canyon in poor visibility

The same basic rules and techniques that apply to flying up a canyon and anticipating possible sink apply equally to anticipating poor visibility. Again, you are primarily concerned with preserving the ability to turn back. Again, a route up against one side of the canyon leaves you room to turn around.

Keep your back door open!

The reason you may have to turn back is that clouds tend to form over the high terrain first. The canyon itself may be in the clear with good visibility, but the pass at the head of the canyon may have closed in with clouds. Now you have to turn back. Usually this is no problem, provided you flew up one side of the canyon prepared to do it; but particularly in springtime, isolated showers in the vicinity can follow you up the canyon, blocking your retreat. Now you really are in trouble. What you should have been doing is constantly checking back over your shoulder to be sure an escape route back down the canyon remains open. As soon as you are unsure about your open escape route, it is time to retreat, even if you did not get far enough up the canyon to see if you could cross the pass and continue on route.

Crossing a pass

Having now successfully flown up a canyon, you are preparing to cross the pass at the head of the canyon and continue on your way. *Everyone* flying in the mountains has approached a pass only to discover he or she could not cross it. This *will* happen to you, too, and no amount of planning or horsepower can prevent it.

It is now at the very head end of the canyon that you should be most alert to the possibility of very unexpected, very abrupt sink and possible turbulence. If you have been climbing into the prevailing wind, it is quite likely that both sides of the canyon as well as the pass itself will have a continuous downdraft feeding into the canyon and

A typical late spring day in the Mosquito Range of the Colorado Rockies, and the author needs to get to the other side to get home. It looks pretty bad in this photograph, but we are flying parallel to the ridge while looking for a window. By flying parallel to the ridge, we have a long time to study things before committing to crossing over. In this case, although you can't see them in the photograph, the Salida, Buena Vista, and Leadville, Colorado, airports are a very short distance away just over the ridge, making it more prudent to take a closer look. Fletcher Anderson photo.

flowing out the bottom. Even if you have been climbing in light lift, that could and often will change abruptly to sink. If there is an anomalous reverse-direction wind, you will probably not feel its effects until you are quite close to the pass. If you should encounter strong sink, you would very much like to be turned around *already* and headed back down the canyon.

The strategy you should employ is to start to turn back at about the point where it would be necessary to turn if you were not going to cross the pass, *whether you encounter sink or not.* Then fly toward the pass on about a 45° heading rather than directly toward it. Now if you have to suddenly turn back, it will only require a 90° turn rather than a full 180° turn.

Once you are right over the pass, duck across and fly directly away from it, down the next valley. This way you are flying out of the

Late spring, Independence Pass, Colorado. Can we continue our flight to Aspen on the other side of this pass? We know it is the right pass despite the poor visibility because no other nearby pass has the two lakes at the foot of the pass or the highway. We plan to take a closer look by flying up one side of the valley while keeping open the option of turning around. On this day we did not find the window at the top of the pass we needed and did not cross. Fletcher Anderson photo.

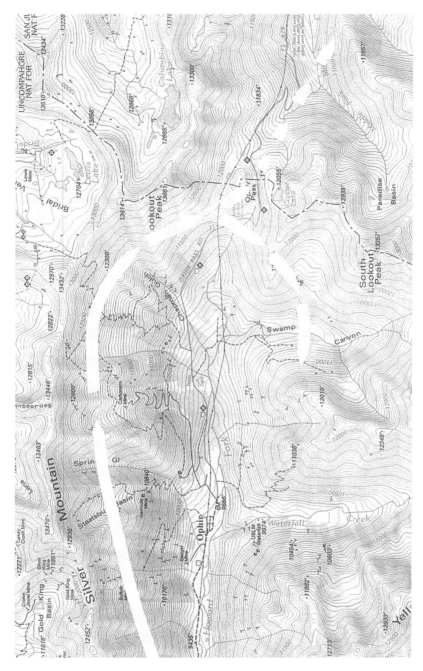

Ophir Pass, Colorado. By flying up one side of the valley, you have plenty of room to make the turn if you need to turn back—this particular valley has a width nearly 3 times the turn diameter of most small planes in a 45° banked turn. At the point where you would have to turn back if you couldn't cross the pass, you begin the turn. That way when you are right at the pass, very little additional turning is required if you are forced to turn back, and in poor visibility you have a longer time to take a good long look into the next valley.

potential downdraft and turbulence on that side of the pass, already headed for lower terrain down the valley.

Bad visibility at a pass

Let us assume you are flying up a narrow valley under a low ceiling. You have kept your back door open; you have left room to turn around if you have to. There appears to be a window over the pass, but what are the conditions on the far side? Is it going to be wise to cross the pass into the next valley?

Why not take a good, long look before crossing over? The same strategy you employed to allow for possible downdrafts will work for you in this instance, too. Start the turn as though you intend to turn back. Halfway through turning around you should be almost right over the pass with an opportunity to take a long look into the next valley. Was it just a sucker hole into a little doughnut of clear air? Turn back. Does it appear the next valley is pretty much in the clear? Are there ways out if it isn't? Is there an airport you can see your way to in the next valley? Then go ahead and cross the pass.

Things don't look too good? No problem—just continue on around the turn which you are already halfway through, and fly back down the valley you came up.

Crossing the Sierras eastbound out of Placerville, California, in the snowy wet spring of 2001, we were not about to go on instruments in a new and unfamiliar aircraft in ice. We had to use this technique in no less than a dozen valleys before eventually finding a way through nearly 150 mi to the south. It was a once-in-a-lifetime view of unparalleled scenery in fantastic light under the clouds rather than an exercise in white knuckles, because the sky to the west was clear *and we always left plenty of room to easily turn back.*

Crossing a ridge or mountain range

This is just like crossing a mountain pass only easier, because you have a lot more room.

Everyone who has ever flown in the mountains has found herself or himself unable to get high enough to cross a ridge ahead and has had to turn back. This is not (or should not be) an emergency. It is an everyday occurrence.

A 10-Knot Wind Over the Mountains

Produces light rotor turbulence. Turbulence is moderate within 500 to 1000 feet of ridge top elevation or less, and generally felt up to two ridge heights downwind.

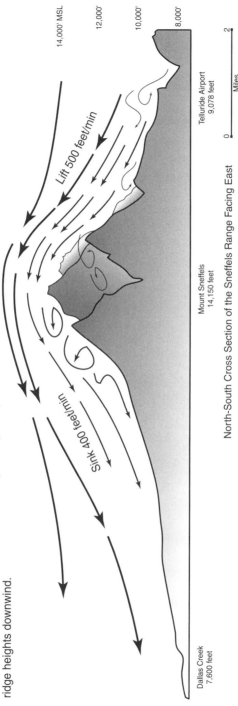

14,000' MSL

12,000'

10,000'

8,000'

Lift 500 feet/min

Sink 400 feet/min

Dallas Creek
7,600 feet

Mount Sneffels
14,150 feet

Telluride Airport
9,078 feet

North-South Cross Section of the Sneffels Range Facing East

0   Miles   2

189

A 25-Knot Wind Over the Mountains

Note the potentially lethal turbulence and rotor. Rotor effect may be felt up to twice the ridge height and over four times ridge height downwind.

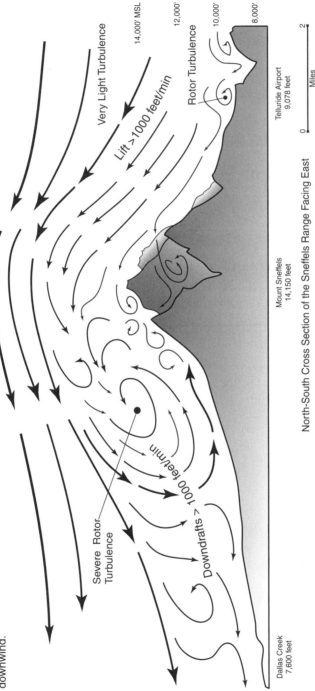

North-South Cross Section of the Sneffels Range Facing East

Why you might need to turn back crossing a ridge: A 10-knot wind over this ridge produces sink rates which exceed the climbing ability of most small aircraft at these altitudes. A 25-knot wind adds severe rotor turbulence. You cannot know with absolute certainty whether these conditions exist until you fly into them.

Approaching the ridge, you should be concerned about the possibility of a sudden loss of altitude. At or well before the point where you would have to start your turn to turn back, turn and approach the ridge on a heading 45° to, rather than perpendicular to, the ridge. Again, if you start losing altitude, you are already partway around the turn that is your escape, so completing that turn should be easier and less dramatic than being forced to crank around in a steep banked 180° turn.

Consider that most of these small-scale phenomena are related to large-scale phenomena—crossing a 200-ft-high, steep ridge involves the same tactics and techniques as crossing the 10,000-ft-high first ridge of the Front Range of the Rocky Mountains climbing out of Denver, Colorado.

Altitude needed to cross a pass

How high do you need to be to cross a ridge or a pass? This is one of those unanswerable questions like "How blue is the sky?" The answer is that you need to be high enough not to be caught in the sink and turbulence on the lee side of the ridge. You don't know how high that is until you actually fly into it. *That means that the rest of the advice in this section is full of assumptions which may not apply to the particular day you have to cross the ridge.* Basic rules of thumb always have exceptions. Nevertheless, these are the rules:

On a day with light winds aloft when you have not been encountering much in the way of turbulence or any significant amounts of lift or sink, at least 1000 ft of clearance over the top of a pass or ridge is probably a good rule. Some people are going to take exception to this and argue that 2000 ft is better. I won't dispute that. More is always better in this instance.

If there *really* is no lift or sink or noticeable wind, you could use the procedures just discussed and cross the pass or ridge at 500 ft, 100 ft, or perhaps even less if you are truly young and invulnerable. The real reason not to is the possibility of unpredicted or anomalous reverse-direction wind. These winds generally, but not always, are confined to within 1000 ft of the ridge top. You need room to deal with not just the sink the winds generate, but also the possible near-stall result of suddenly flying through the shear layer between the prevailing and the reverse-direction winds.

For that matter, there are invariably so many factors affecting wind direction and lift or sink over a pass that it is inevitable that you will misread the indicators many times in your flying career. When you do so, you will be thankful for all the extra altitude you have.

When the winds are strong, rotor on the downwind side of a ridge or pass is almost a certainty. You might be able to descend through the rotor, even though you might not enjoy the ride. In a small, low-powered airplane, climbing through it might be impossible. The altitude you need would ideally have you above the rotor. Unfortunately, it is very possible that your airplane is not capable of getting as high as you want to be. As a guess, though, rotors typically are not felt significantly at more than about twice ridge height. Thus, if the ridge you are crossing is 2000 ft higher than the valley, you would like to be crossing 2000 ft higher than the ridge. Normally you can do that.

In flying toward the Front Range of the Rocky Mountains out of Denver into a 50-knot wind, though, the 14,000-ft mountains loom 9000 ft higher than Denver at 5000 ft. You *wish* you could get up to 9000 ft higher than Loveland Pass, but that would be 23,000 ft and your airplane likely can't do it. On a day like that, however high you can get is better than anywhere lower than ridge top elevation. Once you are beyond the first ridge, the ride may get better, if not ever really comfortable. Then again, you might just try this flight on another day.

A day I guessed wrong

Let me return to a story I have already alluded to. In the fall of 1997 I headed east out of Denver over Loveland Pass toward Glenwood Springs in a Cessna 172. Winds aloft were predicted to be westerly at around 15 knots. Certainly we were encountering no more than that as we flew by the summit of Mount Evans at over 15,500 ft. There was no significant turbulence on the lee side of Mount Evans. The ridge above the Eisenhower Tunnel is only about 12,000 ft msl, giving us a healthy 3500 ft of extra altitude to clear. Remember that a good rule of thumb is 2000 ft? And in less time than it takes to tell it, everything loose in the plane was banging off the roof and sides of the cockpit, the rate of climb indicator was pegged down at 2000 ft/min, and we were headed back away from the mountains.

What is the explanation for that extreme downdraft? Unpredicted mountain wave? Chinook? I still don't have a clue, *but the escape was easy.* We were approaching at a 45° angle to the ridge. It required only a very little turn to get out of there.

The weather briefing I received 15 min before takeoff would not have predicted anything like the weather we encountered only 20 mi west of Denver. *Every pilot who has ever flown a small aircraft in the mountains has at least one very similar story to tell.*

Too late to turn back?

You hope the time never comes when you have flown too far up a canyon or too close to a ridge to turn back. Realizing how serious the consequences of this mistake are should be enough to remind you again that one of the fundamental skills of mountain flying is the ability to convince yourself not to try things you are only pretty sure you can actually get away with. The whole purpose of this book is to provide you with the judgment you need to never find yourself in this situation, because there is no longer a pleasant alternative. Once you are too far into the canyon to turn back, one of three very bad things is going to happen.

Most likely, you will gradually increase back pressure on the yoke in a futile attempt to keep the plane in the air until you stall or stall and spin and crash into the ground. Remember the videotape Dale Wood recovered at the beginning of this chapter?

Next most likely is that things will happen so very quickly, or that your attempted turn will simply not be of tight enough radius, that you will implant yourself into a cliff in full flight without stalling.

Either of these two options is likely to be fatal.

The only thing to try at this point which might save you is a nearly straight-ahead precautionary landing. Is there a flat spot ahead of you? If you are very lucky, you might manage to set down there.

In 1960 my family was returning from skiing over Loveland Pass in a ground blizzard. Larry Jump was skiing slowly ahead of the car so that we would not inadvertently drive over the edge when he heard a large aircraft pass close overhead and then heard both its engines abruptly stop. Larry and my father quickly skied off in the direction of the last sounds of flight. The aircraft was a military light twin-engine transport. (A number of them were engaged in training exercises near

Camp Hale near Leadville, Colorado, in support of the upcoming Bay of Pigs fiasco.) The pilots had been flying in clouds at 11,000 ft on the decidedly incorrect assumption that that would put them above any terrain. Not so—Loveland Pass itself is 12,000 ft msl; several nearby peaks are over 14,000 ft msl.

While this was a patently stupid error, the pilots' reactions when snow and rocks loomed up directly in front of them saved their lives. They instantly pulled both engines to idle and bellied in in deep powder snow. So soft was the landing that a passenger sitting in back with no seat belt and reading comic books was not even thrown out of his seat. The aircraft had to be removed by truck in the summer, but was repairable.

This might save you, but don't count on it. Survivors of crashes at Independence and Haggerman Passes interviewed by Baker and Lamb or rescued by Richard Arnold reported attempting precaution-ary landings which saved some of, but not all the aircrafts' occupants. Obviously success depends heavily on there being a flat, open space available. Where there are only trees, by the way, your best hope might lie in trying to perform a conventional landing *on top of* the trees into the upper branches. Dick Arnold, when he was a director of mountain rescue in Aspen, Colorado, performed several rescues of people who saved their lives at the last moment with this maneuver.

Making the decision to turn around early is really a much wiser option!

Traffic at mountain passes

You have been taught or have observed that the two places where you are most likely to encounter other traffic are at airports and over VORs. Everybody's routes converge at those points.

Mountain passes are the same way. All the small airplanes in the sky will be planning routes across the mountains which use a relatively small number of mountain passes. Like you, they will fly lower alti-tudes for most of their route and just pop up high enough to safely cross the pass. You know the recommended cruising altitudes are even thousands of feet plus 500 westbound, odd thousands of feet plus 500 eastbound. Do not expect anyone to actually be using these altitudes as they cross passes or ridges! They (and probably you) are going to climb up only as high as needed to safely cross the pass, and then go back down to altitudes where the plane flies better and they don't need oxygen. Even on route, safe terrain and cloud clearance rather than recommended cruising altitudes are going to determine how high everyone flies. Even if they were using recommended altitudes, they would be in the process of climbing up to or descending down from those altitudes, just as you were doing the opposite. I have heard the advice given that you should always fly the right side of the valley, similar to staying in the right-hand lane on the highway. This might be a good idea, but there will be days—most days—when one side of the valley is definitely preferred for flying conditions. *Both ascending and descending pilots will fly that side of the valley if they read the conditions correctly.*

Consider the following example. Several years ago, Terry Sargent was flying a charter north over a pass outbound from Gunnison to Aspen. Who should be headed south at that very moment but Dick Arnold, flying another charter over the same pass from Aspen to Gunnison! Even several years ago, these were two very experienced, very competent, very cautious mountain pilots. Accordingly, each was flying up the east, lifting, side of his respective valley. At the pass, even though each was pretty sure of making it over, each started the turn to the west needed to turn back just in case and began approaching the pass at a very sharp angle to cross. And it is a good thing they did, because whom should Terry spot right off his right wing but Dick, and at the very same moment whom should Dick see but Terry, right at his altitude on a slightly converging course! There were probably only two planes anywhere over the Elk Mountains that day, and by following the very best safety practices they knew, both pilots managed to arrive at the exact same spot simultaneously.

Although almost no one does so, it would be a very good idea to simply announce your intentions over the radio, using the frequency of the nearest airport as you approach a pass just as you do when you come in to land. Using the frequency of the nearest nontowered airport, you would announce, "Blank Pass traffic, Cessna XYZ is headed north over Blank Pass at XXX thousand feet, any traffic please advise." I can count the number of people I have actually heard do this on two fingers: That would be Joline Esposito southbound over McClure Pass en route to her home in Telluride and Murray Cunningham southeastbound over Taylor Pass outbound from his home in Aspen. But I *saw* both of them and probably would not have spotted either one if I had not been alerted by the radio call.

Instrument flight in the mountains

After all this discussion of the problems of encountering all sorts of unexpected clouds and other weather in and close to the mountains, why not just file for IFR and fly on instruments? After all, this is what the big airliners do. You don't find them thrashing through turbulence low in among the peaks.

This question is asked most frequently by flatland pilots who live and regularly fly at lower elevations in very humid climates with lots of instrument meteorological conditions (IMC) days. It is asked by coastal pilots who routinely have to climb out through and descend to land through layers of morning ground fog. Without the ability to fly on instruments, the airplane would have little practical use in their environment, and they usually fly on an instrument flight plan. In fact, they usually do so even on days with perfect visibility because they live in very populated areas and the instrument traffic controllers provide them with traffic separation.

These people's methods make excellent sense *in their particular environment*. I often fly into the crowded Denver, Colorado, class B airspace on an instrument flight plan or at least in contact with Denver Approach Control strictly for traffic separation reasons. This method is indeed the way to fly over the mountains in a large, pressurized multiengine aircraft. For a variety of reasons, actual IFR is not wise or practical for a small aircraft in the mountains.

Experienced IFR pilots will still reasonably ask again, Why not IFR in the mountains? Yes, you can experience bad weather in the mountains, but you have bad weather everywhere in the world. The very nature of instrument flying is inclement weather. If the weather were never bad, instrument flying would never have been invented. Sure, the mountains have their share of weather, but so does the rest of the world. Mountain thunderstorms are nothing compared to those tornado-spawning monsters found over the Great Plains. The mountains may have thunderstorms, but those storms are seldom embedded in an overall regionwide cloud layer. We may have low-pressure systems come through, but ours do not reach the level of a Gulf Coast hurricane which blows away houses.

Why, they ask, would you engage in all these elaborate strategies for flying through the mountains when you could just file IFR and fly high above them, and above the tops of the clouds as well? If your aircraft is capable of flying high enough, that is usually the best thing to do. In fact, if your aircraft were capable of flying that high, you would probably choose to do so on any day, regardless of weather.

Don't

Still, *the best advice for flying IFR in the mountains in a small aircraft is, Don't!* The reasons all relate to altitude, not the rugged terrain.

Let us go over a few of the reasons not to, one by one.

Turbulence

The fact that you are in a cloud and cannot see anymore does not mean that the turbulence and sink you can encounter over the mountains on a good-visibility day will conveniently go away. Now, though, you have to be flying an assigned route at an assigned altitude. You don't know precisely where all the mountains are, and the strategies you normally employ to deal with turbulence and sink are no longer available to you. Severe or extreme turbulence can accompany storms in the mountains. Although this level of turbulence is generally very predictable by Flight Service, a small, low-powered airplane cannot fly high enough to be above it.

Minimum en route altitudes

Minimum en route altitudes are generally 2000 ft above the highest nearby terrain. This means that if you are westbound on instruments out of Denver, Colorado, headed over the mountains, the lowest altitude you could expect to be assigned anywhere is over 16,400 ft, but radar cannot see you when you are that low. On portions of most routes, 18,000 ft msl is the lowest usable IFR altitude, and even that altitude is not available when the barometric pressure is low. Coming back eastbound, you would not get lower than 17,000 ft, and more likely you would be assigned at least 19,000 ft. These altitudes in and of themselves are higher than those that the vast majority of small, low-powered aircraft are capable of flying. Even those small planes capable of getting this high are at the upper limits of their ability, so actually maintaining those altitudes in strong up- and downdrafts would be problematic.

In flying on instruments from Telluride to Denver via the Blue Mesa VOR, there is a 40- or 50-mi-long segment of the flight near Gunnison where you are lost to radar at 16,000 ft msl or lower. Should weather force you below about 15,000 ft, you sometimes also lose radio contact with the controllers for about 10 or 15 min. This is no problem if you are talking to them for flight following in good weather but more serious if you are relying on them in a storm. On this route 18,000 ft msl is required for safe instrument flight, and again, most small aircraft can't do that.

Published departure climbs

Surrounded as they are by sometimes very high mountains very close to the airport, most mountain airports have published instrument departure procedures that require steeper than normal departure climbs. Many of these climbs cannot be done in most small, low-powered aircraft. Remember that your climb rate at these altitudes is about one-half what it is at sea level at the same speed. To name just a few examples, I have repeatedly been unable to maintain the published departure climbs out of Eagle, Rifle, and Aspen, Colorado, with only moderately loaded small aircraft when giving instrument instruction to students under the hood. Had we been in real instrument conditions, we would have had to circle for altitude, *and we would have had to do so at altitudes too low for the controllers to have acquired us on radar and advised us.* The Rockies Four published departure procedure out of Denver, Colorado, requires a sustained climb angle up to 18,000 ft that virtually no single-engine plane, turbocharged or not, can maintain.

Missed-approach procedures

As with published departure procedures, published missed-approach procedures can require climb angles which small, low-powered aircraft cannot execute at high mountain altitudes.

Approach procedures

Instrument approach procedures into mountain airports for the most part are no more complex or difficult to fly than those at any other airports, with one notable difference. Descents can sometimes be much steeper than normal in order to clear nearby terrain. For a small plane, a steeper than normal descent is not really a problem—most small planes suffer from plenty of drag and can easily keep their speed under control on a steep descent. Terrain is much closer to the approach path than is the case for a flat land approach.

Minimum descent altitudes

To stay clear of terrain, missed-approach points and minimum descent altitudes (MDAs) are often very high. The MDA of Aspen, Colorado, is 2500 ft above the runway; Telluride's is 2200 ft for the localizer DME approach, but over 3000 ft agl for the VOR approach; that of Eagle, Colorado, is 2700 ft above the runway. Are you beginning to see the picture here? You can actually miss the approach when the airport itself down below you under the clouds is in fine VFR conditions! There are a lot of days when you will have to miss the approach and go to an alternate airport. Since these alternate fields are not always very closeby, you will have to be carrying plenty of fuel, and that may reduce both your useful load and your performance below practical limits.

The very high minimum descent altitude at Aspen was a contributing factor in the nighttime crash of a Gulfstream jet in the winter of 2000–2001. The pilots had the runway in sight at the missed-approach point and turned onto final approach. Then in the dark they flew into another cloud and lost visual reference with the ground. They were now too far below nearby terrain to attempt the missed-approach procedure, but could not see the runway to land either. They flew into a steep hillside just to the right side of the approach end of the runway.

Here is an interesting thing people do at two of the last three mentioned airports: Missed the approach at Aspen? Go back to the Red table VOR and fly the instrument landing system (ILS) into Rifle, Colorado; then fly almost 50 mi back up the valley VFR under a very low ceiling all the way back to Aspen. During the busy Christmas season, Twin Otters hauling freight often do this. Can't get into Telluride? Go back to the Cones VOR, then fly the ILS into Montrose, Colorado, and scud-run 30 min back under the ceiling to Telluride. Please remember, no one claims that this is a particularly good or safe idea, only that it has been done in the right conditions.

Mountain icing

Icing is the most important of all the reasons not to fly IFR in the mountains in a small aircraft. No small aircraft really carries ice very well. The amount of ice buildup is related to a degree to aircraft speed, but not to size. All else being equal, the smaller your wing, the more the same amount of ice disrupts the airflow. An inch of ice at the leading edge of a wing with a 3-ft chord is a bigger percentage of airfoil shape than the same inch of ice is on a 6-ft wing. Also, very few small aircraft have deicing.

Because of the altitudes involved, icing conditions can be encountered over the mountains at any time of year. Weather prediction over the mountains is notoriously poor; weather observation points are few and widely scattered. Few people are flying on instruments, so pilot's reports are scarce. The basic ingredients required for icing are visible moisture in the air and temperatures below freezing. At altitudes over 14,000 ft msl, that can occur any day of the year.

Any time of year? Let's take as an example July 5, 2001. Despite weather forecast to be near-perfect VFR the whole way, 10 min after takeoff from Hood River, Oregon, we were in the clouds. Minimum en route altitudes over the mountains of Idaho and Utah had us taking mild ice soon thereafter, and for the next 4 h we were constantly changing altitude to get out of ice. Finally beyond the clouds, we descended into Moab, Utah, for fuel, where the ground air temperature was…105°F!

Whenever you encounter icing, no matter how mild it seems at first, you have to do something about it right away. In a small, low-powered aircraft at high altitude, there may not be any options available.

Normally the icing layer is only about 3000 ft thick. That means if you took off from a sea-level airport and encountered icing at, say, 4000 ft, you could just climb up above 7000 ft or descend back below 4000 ft. This is no problem. Almost any small aircraft can do that. At those altitudes a climb rate of 700 ft/min could be expected, so you would only be in icing conditions for 4 min. (Okay, we all know there are plenty of days when the icing is too severe to do this.)

Corporate jets and larger turboprops just climb rapidly through the ice. In the mountains in a small, low-powered plane, if you encountered icing on the climb-out, chances are very good that you would not be capable of climbing high enough above the minimum en route altitude to get above the icing layer. Neither could you descend below the ice because the minimum en route altitudes are so high. In a very short time, you could be in big trouble. The minimum en route altitudes over most of the Colorado Rockies are all at least 17,000 ft msl. That means a climb to over 20,000 ft msl would be a likely requirement. Can your plane do that? Can it do it once it is already carrying ice? How long would the climb take once ice reduced your already very slow climb rate? How much more ice would you accumulate?

On a descent if you encountered ice, you might pick up enough ice that you could not climb again if you missed the approach. Remember that at these altitudes your climb *without* ice is only one-half what you get at sea level.

Again, remember that because of the altitudes involved, you can encounter icing conditions over the mountains in any month of the year.

While "normal" kinds of icing conditions are trouble enough, two varieties of icing unique to the mountains can overwhelm any small aircraft.

Upslope icing.   When an air mass is blown against the mountain front, it is deflected upward by the terrain. As the air is lifted, it cools. Cooler air holds less moisture, so some moisture condenses out of the air. The upwind side of a mountain range can be entirely encased in cumulus clouds which will tower far above the mountain summits. Now you have below-freezing temperatures and visible moisture. On a bad day, when you have icing temperatures beginning at, say, 6000 ft the icing conditions themselves can be occurring in the lifting air, and because the air mass is so rapidly lifted, icing can persist all the way up to and above 18,000 ft. This is so because small droplets of moisture will "supercool" to well below the freezing point but will not yet have frozen. Normally, $-20°C$ is the temperature below which icing no longer occurs. Supercooled droplets can be encountered at temperatures far lower than that.

On a very bad day, when droplets of moisture in the air become supercooled as they are lifted in the rapidly cooling air, they remain as supercooled liquid because they were lifted so very rapidly to such high altitudes. For some reason, they can persist at temperatures as low as $-40°F$. In rapidly lifting air, more supercooled droplets are constantly replacing the existing ones, whereas in more stabile air, one by one they gradually freeze out of the system. These droplets are just waiting for almost anything to disturb them. When it does, they freeze instantly. They will freeze absolutely instantaneously if they hit any object cooled below the freezing point—such as your wing.

Here is a typical mild encounter with upslope ice: On a descent into Telluride in the spring in a Cessna P210, ice was not forecast, but we began taking on light rime ice at 22,000 ft msl. We had already been cleared to descend at our discretion, so we elected to get down

UPSLOPE ICING

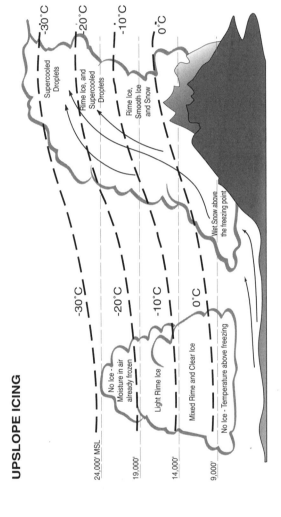

24,000' MSL	-30°C
	No Ice - Moisture in air already frozen
19,000'	-20°C
	Light Rime Ice
14,000'	-10°C
	Mixed Rime and Clear Ice
9,000'	0°C
	No Ice - Temperature above freezing

-30°C

Supercooled Droplets

-20°C

Rime Ice, and Supercooled Droplets

-10°C

Rime Ice, Smooth Ice and Snow

0°C

Wet Snow above the freezing point

Normal Ice
In this hypothetical cloud, you would encounter ice between 12,000 feet and 15,000 feet MSL, but most of that 3000-foot band would be light.

Upslope Ice
In this cloud, strong upslope winds are lifting unfrozen water droplets from below 9000 feet to 24,000 feet MSL, where they remain supercooled and unfrozen. You could encounter supercooled droplets from 12,000 feet up to over 20,000 feet. Below 9000 feet, wet snow impacting could still adhere to the airframe.

through it as quickly as possible, which for us was a descent rate of 2000 ft/min. We were not accumulating very much ice, *but we continued to take on more ice until we broke out of the clouds at 11,800 ft.* At 11,000 ft we were finally in air warmer than the freezing point and began shedding the ice. This could almost be called a typical spring day. When we broke out of the clouds, we were carrying less than half an inch of ice, and that in a reasonably smooth layer right on the leading edge of the wing. This in itself is no problem for that aircraft. Flight characteristics are not noticeably affected at all, but had we missed the approach, we would have been flying in ice for the next 45 min to our alternate airport. At the same rate of accumulation, that could have been a problem. *The icing layer was 11,000 ft thick, rather than the more normal 3000 ft thick.*

Supercooled droplets. Stronger lift keeps very large supercooled droplets aloft, so the rate at which the ice builds up can be dramatic. Smooth ice will quickly envelop you. Icing from supercooled droplets can be heaviest in the highest parts of the clouds, and owing to temperature fluctuations rime ice can add to the smooth ice at other elevations in the same cloud.

This brings up our next example. The same Cessna P210 as before, same location two weeks later, was flown this time by my neighbor Todd Wilson. The weather prediction was better than before, and in actuality the weather appeared to be slightly better than predicted, but spring weather in the mountains can best be described as volatile. The previous aircraft to fly the approach reported no icing. Todd was in and out of clouds at 20,000 ft. Seconds later Todd gave his passengers a very concise but extremely comprehensive briefing. His exact words were, "Oh, shit!" In scarcely 30 s he had taken on 1/4 in of ice. Passengers described it as though a huge bucket of water were thrown on them and instantly froze. That was the first encounter with the supercooled droplets.

Does 1/4 in of ice warrant this reaction? A Cessna 210 might ultimately be capable of carrying considerably more, and flying night freight over the mountains meant that Todd had no lack of familiarity with icing. Todd had already figured out that very immediate action was going to be required, and he was running through his options. Delaying too long would make a climb back above the ice difficult. The 45-min diversion to his alternate airport might or might not become equally demanding. Among other options, he was considering the possibility of using his GPS and asking for vectors to an airport farther west than his alternate, which he knew was in the clear because he saw it from above the clouds, but an airport with no published instrument approach. The situation was far from out of control and far from life-threatening, but his options were quickly running out. Certainly he was not in trouble yet, but Todd was already thinking way ahead and planning accordingly. In a situation like this, you want to be flying with a pilot with the presence of mind that Todd routinely displays.

This initial icing was smooth ice in the form of supercooled droplets, and Todd descended below that layer as fast as he could. He was out of the clouds, into the sunshine, then back into the next cloud. Whap! There was another blast of supercooled droplets. He was back in the sun between clouds. Smash! The next cloud was full of graupel (which is a form of pellets of snow quite common in spring). Now he was back in the sun for a very brief instant. Todd and his passengers also report encountering at least one cloud full of hail, two kinds of snow, and plenty of relatively light rime ice which kept accumulating. Icing was diminished at lower altitude, and in the form of light rime ice, but still accumulating.

All were pretty much in agreement that they had really already derived the full entertainment value from that flight and were ready to go home now—which they were not going to be allowed to do! Since another aircraft had just departed Telluride on instruments, the traffic controller could not clear Todd for the approach into the same airport and stuck him in a hold 15 mi out at the nearest VOR.

In the clouds, out of the clouds, in again, out again—Todd could clearly see that this was not a good long-term proposition. Fortunately slightly west of the mountain front where he was holding, the clouds and snow showers were much more widely scattered. In the middle of the biggest gap between clouds out there he canceled IFR and kept circling in the sun between the clouds. There he began shedding ice, and once the departing traffic was behind him, he made a visual approach into the runway, doing cloud slalom and staying in clear air.

By the time he broke out of the clouds for the last time at about 13,000 ft, he was carrying more than 1 in, if not 1.5 in of ice and gradually taking more. Worse, the mixture of icing types had formed the ram's horn configuration, which is most disruptive to airflow (see illustration).

By staying well ahead of the situation and considering several options, Todd dealt very well with a potentially very dangerous situation.

The next aircraft to land after Todd was a Beech 1900 airliner, flown, as luck would have it, by Great Lakes Air's then senior captain, Jeff Garcia. Jeff came straight in on the localizer instrument approach. Although that airliner has extensive deicing capability, the winglets are not deiced and were carrying nearly 3 in of ice.

In the first 30 s of an icing encounter, on a very very bad, really horrible day, the upper reaches of this hypothetical hunk of moisture-laden, rapidly cooled air will be like the description of the previous paragraphs, forming rime ice (the white opaque stuff) *as well as* containing supercooled droplets and forming smooth, clear ice. This mixed ice can be very dangerous, as it accumulates very quickly and forms a very irregular surface. This is what happened to Todd. But on a day when you wish you had stayed home in bed, the very lowest reaches of the *same clouds* would include very sticky wet snow. Your aircraft would be accumulating big chunks of smooth ice from the supercooled droplets mixed with the irregularities of the rime ice. With exceptionally bad karma, you would next be pasted with plenty of wet, sticky spring snow, which would be more than enough of a problem all by itself. Anticipating this problem, Todd took evasive action before it could occur.

This is the combination that is the most dangerous of all icing. With upslope wind conditions, all varieties of icing can be encountered over a very large range of altitudes. You could be taking on ice all the way from pattern altitude up to over 20,000 ft. This is far from a typical day, and you would probably realize that the icing was too extreme to mess with before you took off. Few experienced mountain pilots would knowingly fly a small aircraft with no deicing capability into a wet snowstorm. Snow impacting on the airframe is its own special category of icing.

But human nature being what it is, someone is always willing to risk it anyway. And this leads us to our third example.

Snow. It's the same location, one more week later. The instant he encountered ice, Todd's reaction was automatic. How do I get out of this as quickly as possible? The examples themselves tell you why that is the only appropriate response. There is simply no sure way to tell ahead of time how much more ice you will encounter.

But not everyone thinks as Todd does. For example, a week later in the same spring, a freelance flight instructor elected to take one of his students into ice and a fairly heavy spring snowstorm *deliberately*, in order to teach him how to handle it. This is not quite as reckless a procedure as it might seem to be *if the aircraft has deicing capability, if the icing is reported to be light, if the weather is improving, and if there is an obvious way out.* In a well-deiced aircraft, you might elect to climb up into a well-defined icing layer for the purposes of instruction. None of those conditions were met, and weather can change very quickly in the mountains.

They simply climbed up into a cloud with reported icing conditions west of Telluride. Previous aircraft in the same cloud had reported light to moderate mixed icing. The idea was to go up into the cloud for just a little bit, then drop back down out of it.

Sure enough, the reports were correct, and the Cessna 182 began gradually accumulating ice. The Cessna P210 in the previous two examples has a heated propeller and a hot patch on the windshield, so its ice was only on the airframe. This was not so with the Cessna 182, which began accumulating small amounts of ice on the propeller as well. The vagaries of spring weather being what they are, things rapidly changed and they began accumulating lots of impacting snow as well as ice. A Cessna 182 does have a heated pitot tube and an alternate static air source. Perhaps neither was turned on; or perhaps the static port, the pitot tube, or both froze over anyway. In any event, airspeed, rate of climb, and altimeter ceased to function properly. Both communications antennae were encased in ice, and one broke off. The pilots now did not know exactly how high they were or how fast they were going, and they could not talk to anyone. In fact, they were going down very fast and broke out of the clouds in a steep dive. Although they broke out at a reasonable altitude above the ground, the cloud base had lowered considerably with the falling snow. Although they were only in the clouds for a couple of minutes, the outside air temperature was still below freezing under the clouds; so even though they were now in clear air, they were not shedding very much of the ice they had acquired. The propeller was still carrying enough ice to be well out of balance. It was shaking badly enough to damage the propeller seal and cause a significant oil leak and to severely tweak one of the aging motor mounts. The combination of an undetected dive above the aircraft's maneuvering speed and an abrupt pull-up combined with airframe ice had damaged already worn elevator bushings. The loose bushings and ice on the stabilizer caused intermittent elevator flutter, making the damage progressively worse.

Presumably the aerodynamics of the plane were not that optimal anymore, but with no more ice accumulating there was no immediate difficulty to keep it flying. Normally this would be the biggest issue with icing, but in this particular instance it was not. The big concern is that ice buildup on and above the leading edge of the wing destroys the airfoil, and you stall. When this will or will not happen depends on both the thickness and the roughness of the ice, and it is not very predictable.

Imagine the consequences if ice continued to accumulate, the iced up pitot and static ports were not discovered, and the descent became a little steeper and faster, the ice thicker, or simply the cloud base lower so that they did not break out. For that matter, imagine the consequences of having to continue the flight much longer with the out-of-balance propeller or fluttering elevator alone. Engine damage was already occurring. At some point the elevator could have eventually parted company with the aircraft.

Still not frightened? The repair bill for this lesson was reported second-hand as $10,000. Probably this figure has been exaggerated because while the propeller required a complete overhaul, the engine needed a new propeller seal and one new motor mount, and the stabilizer needed new bushings, the engine did not get a teardown.

This instructor is still teaching people how to fly in ice. Did this repair bill sober him? Not really—it was the *student's* airplane!

How well can icing be forecast? The next aircraft in following this nearly fatal incident was another Beech 1900 airliner, this time flown by Troy Beatty (on one of his very first airline flights to Telluride and who eventually was promoted to company check airman), who was carrying…no ice whatsoever. He had maneuvered around the worst cloud, rather as Todd Wilson did in the previous example.

Mountain wave icing

The updraft portion of a mountain wave can produce conditions similar to, but even more severe than, conditions found with upslope icing. The sheer magnitude of air movement encountered in mountain wave can greatly aggravate these conditions.

To the essential ingredients for mountain wave, high winds perpendicular to the mountain range add visible moisture at the wrong temperatures.

Go back to the discussion of upslope icing, except that the rate of climb in the upwind portion of mountain wave can have sustained climb rates well in excess of 1000 ft/min going up several thousand feet. This not only favors the propagation of supercooled water droplets, but also may see those droplets still in liquid form when the temperature has dropped way down to −30°F. These are just dying to adhere to your airframe.

This may be of less concern to small aircraft only because these conditions are more often found at much higher altitudes where smaller aircraft do not venture. Mountain wave icing is typically a summer phenomenon—in winter the moisture at those altitudes is often already frozen and doesn't adhere to anything.

Single-engine IFR in the mountains

You have just been reading about it, so people must do it. Despite the fact that it is dangerous, the newspapers are not full of accounts of such crashes. Yes, it is possible, and it is done regularly. Frequently I find myself doing it every week during the winter months. It is no harder to do than with IFR anywhere else, but your margin of safety is far less than IFR elsewhere. If you intend to try this, you really should be in an aircraft capable of easily exceeding 22,000 ft to get safely above the icing layer, if not the weather, and you should have deicing. Even then, certain varieties of mountain icing conditions mean that you should avoid days with predicted icing conditions even if your aircraft has deicing. You should absolutely forget the idea of flight in wet, falling snow in any small, single-engine aircraft.

In a typical four-seat, low-powered, small aircraft, performance is insufficient to meet the combined demands of altitude and ice. Despite the fact that I fly a Cessna 172 year-round as part of my job, I typically accumulate less than an hour a year of actual IFR in the mountains in that kind of aircraft, and neither does anyone else. Given that I fly on a schedule weekly in a Cessna P210, the total is still something less than 20 h/yr of *actual* instrument conditions, although probably well more than double that number could be termed, but not logged, "instrument flight" because the entire flight may be on an instrument flight plan, even though only the climb and descent are actually in the clouds. In a higher-performance, small aircraft, you can do some mountain instrument flying under some conditions. The ultimate truth is that small, single-engine aircraft are not all-weather aircraft.

Even then, the best advice is wait for a better day.

Psychologically, when flying VFR, you are entirely the master of your own fate. The very instant you feel the need to do something differently, you do it. Flying on instruments, you are at the mercy of a controller 200 mi away in a dark room. Perhaps saying *at the mercy of* is unfair. Purely and simply, traffic controllers are there for one purpose only, and that is to help you. (After all, you are not up there flying around for the purpose of helping them.) Controllers provide you with a tremendous amount of help every time you contact them.

Nevertheless, the instrument controller cannot be expected to instantaneously evaluate your circumstances, and because she or he is simultaneously dealing with several other aircrafts' needs, positions, and assigned routes, the controller cannot always give you a clearance to do what you need to do as quickly as you want it.

An IFR strategy to avoid

There is one local pilot who flies someone else's small aircraft who always seems to be able to get in under IFR conditions even when everyone else, including the airlines, can't do it. How does he do it? Presumably everyone flies the same published approach and encounters the same visibility at the missed-approach point, yet this guy gets in anyway. His fans, and he has many, claim he is just a better pilot.

The secret procedure is this: The air traffic controllers at Denver Center control Telluride's instrument approach, but they cannot see the whole approach on radar. (This same situation is true at many mountain airports.) Beyond a certain point on the approach they lose sight of the craft. At that point, the savvy pilot simply drops down another 1000 ft lower than the published minimum altitudes. Occasionally he drops down to 500 ft above runway elevation compared to the authorized 2200 ft. Then the pilot continues inbound right to the end of the runway instead of missing the approach at the missed-approach point 1 mi out.

The pilot can see the runway when everyone else can't because he is looking for it from 1000 to 1500 ft lower and 1 mi closer. Provided that the ceiling is high enough (and ATIS has that information and the pilot checks it before coming in), this is no more difficult than flying the legal approach procedure. In fact, thanks in part to an autopilot linked to both the GPS and the Localizer, the procedure itself not only is no more difficult than the legal approach, but also is done exactly the same way. It requires no special skill, only a more flexible approach to regulations. The autopilot holds altitude and tracks courses with a much higher degree of precision than the procedure requires.

Why don't the regulations just allow everyone to do this? Why don't the airline pilots do this? Are they too dumb to figure it out, too chicken to try, or just such regulation nerds that they won't consider anything different?

They can be kind of nerdy about regulations sometimes, and you should be, too. There are two reasons. Of less importance is the fact that minimum instrument altitudes are set to keep you above terrain within specified distances of the approach path. That way, if you are a little bit off, you still don't hit anything. No doubt you can fly approaches with a very high degree of precision, but have you ever had anything go wrong? After all, airplanes are made by humans, and anything that people make, can break. In one particularly trying week during the spring of 2002 in three different aircraft, I had three different problems on three consecutive approaches which made my flying much less precise. An autopilot shifted to reverse sensing on an ILS into Jeffco/Denver, Colorado, forcing a missed approach in ice and a second attempt right down to minimums. An attitude gyro tumbled on an NDB approach into Spearfish, South Dakota, making a nonprecision approach very nonprecise. Finally, a vacuum pump failed on an ILS into Montrose, Colorado; but by then I was getting reasonably current on partial panel approaches. This was a fairly trying week. They say problems come in threes, so at least now (to mix metaphors) I don't have to wait for the other shoe to drop.

More important than a lack of precision, if you were to try this secret technique and then had to miss the approach, you would have to execute a 180° right turn back out of the valley. The day may come when reported visibility at the runway suddenly changes in a snow flurry just as this pilot is nearly there using his secret approach. This would have him (or you, or the airline pilots) aimed right at a very large hill in the middle of the valley which is considerably higher than the runway. None of your instruments can tell you where that hill is. Your climb might or might not get you above it.

The pilot who does this has said that it makes him a safer instrument pilot, and he can do it because of his higher level of skill. No doubt *when this works,* it *is* safer than continuing to fly around in the clouds, and it is certainly much more convenient than diverting elsewhere. The day it doesn't work may be fatal, and this pilot regularly carries children. Odds are that day will never come, but why take the chance?

Regulations do not precisely fit every situation all the time, and no doubt all of us are sometimes smarter than whoever devised each regulation (or at least we think we are). But the FAA considered a great many things when it established the published procedures, including an exhaustive analysis of historic accidents, and without knowing what all those things are, we do better not to second-guess them.

5

Specific mountain weather concerns

Having decided to leave instrument flight in a small aircraft in the mountains alone, but still feeling the urge to fly in marginal weather, we examine two possible alternatives:

Flight above the clouds
VFR on top unlikely in summer

If flying through the clouds in the mountains is an inherently dangerous idea, perhaps simply flying on top of them might be better? Yes it is. Use of VFR on top of the clouds over the mountains is quite possibly the most comfortable way to fly bad weather, *provided a number of conditions are met*. Most obviously, you need an aircraft capable of flying high enough to make that option possible. Over high mountains, that can be very high indeed. In December through early April over the mountains of Colorado, cloud tops of 17,000 to 18,000 ft are quite typical. Cloud tops of only 14,000 ft are not too unusual. Cloud tops much above 20,000 ft are uncommon. A typical commuter airliner or corporate jet will always be on top in the winter months. A turbocharged, high-performance, single-engine aircraft usually can be. What this means is that flying visually above the clouds can very likely mean flying above 18,000 ft. Most small aircraft can't do that. Above 18,000 ft, by regulation you are on an instrument flight plan, and you need oxygen. Even flying IFR, you would rather be out of the clouds, out of potential ice, and comfortably in the sun. Usually the traffic controllers can help with that.

There is a profound difference between flying *in* the mountains and just flying *over* them. In Colorado, for example, you could leave

A typically confusing mountain weather picture. Lenticular clouds above suggest wave. A cloud waterfall in the foreground below suggests possible turbulence and a pressure drainage wind. By staying high, these pilots are avoiding the turbulence. John Kounis photo.

Grand Junction or Montrose, which lie in open country 50 mi or so west of the central mountains, cross at high altitude, and descend into Denver, 30 mi east of the mountains, and never have to deal with clouds at all! Since weather tends to develop over the mountains, much of the time the weather at either end of the flight would be wide-open VFR. Great! (If that is in fact the way the weather works out. Weather has a remarkable way of not doing what you need it to do on the day you need it.)

This is an excellent idea *during the winter,* but various factors must be taken into account. Remembering that weather prediction is notoriously poor over the mountains, you could easily try to cross the mountains on top of the weather, only to discover that cloud tops were gradually rising to altitudes which your small aircraft could not attain. In most small, low-powered aircraft, the minimum instrument en route altitude could be higher than you could fly, leaving you in the clouds, off the radar, out of communication with traffic controllers, and closer to the mountains in the clouds than you would like to be.

Rapidly changing mountain weather can and frequently does mean that weather that was reported as clear blue skies when you began the flight has changed to IFR before you reached your destination. The flight in clear air on top for the en route portion of the flight is certainly better than grinding along through the clouds the whole way; but you must be prepared to make an instrument approach through the clouds at your destination. Although you would take on less ice in a rapid descent than you would in a prolonged cruise through the same clouds, ice is still possible.

Flights beginning at and/or ending at mountain airports are a different proposition. In those cases, all the concerns about IFR in the mountains are relevant. You would be on an IFR flight plan, but in visual conditions above the clouds in order to avoid the possibility of icing during the en route phase of the flight. The climb and the descent would be through the clouds.

Warmer summer air is less stabile. Condensation occurs at higher altitudes, and clouds build more quickly. Summer convective cloud buildup frequently reaches tops of over 24,000 ft, and tops reaching over 30,000 ft are certainly not unusual, even if not a weekly occurrence. With warmer temperatures, cloud tops are almost always substantially higher than your small aircraft will fly. Tops being higher, it is more likely once you are on top to find that cloud tops are lifting at a rate that exceeds your ability to stay above them. If you are headed over the mountains to a destination out in the flatlands, your destination may or may not remain in the clear. A destination in the mountains could be socking in beneath you. Certainly on a typical Rocky Mountain summer afternoon, a descent through clouds to your destination in the mountains includes the possibility of thunderstorms.

What may sometimes work very well for you is visual (VFR) flight *between* widely scattered clouds. This does not mean between cloud layers, which should really be an instrument procedure; rather it means flying around and among the clouds. The ride may be noticeably smoother than bashing along beneath the clouds. You maintain good ground reference as long as the clouds remain far enough apart. Remember, if you do this, that aircraft flying on instruments may be flying through these same clouds, which is why the FAA established cloud clearance as well as visibility requirements.

Flight above low stratus clouds

Flights above low stratus clouds in the valley bottoms present few challenges. Clouds of this type are typically formed of residual moisture from a previous day's precipitation. As with any other situation flying in the mountains, be sure you have a way out. Ideally, you would just be crossing over low clouds in one valley and be able to see where you could easily reach the edge of the cloud bank. These clouds are typically markers of very stabile air. The cloud layer marks the top of a temperature inversion which formed the previous night. Cloud tops are seldom rising, and the morning sun will burn them off.

A word of caution, though. On a day when there is weather on the way, for example, when a cold front is coming, as the air becomes unstable, these very stabile low stratus clouds can begin to transform to scattered building cumulus. Many people have been surprised in these conditions. Few have been trapped—the initial buildup is slow enough, and the tops low enough, to allow you to find a way down safely.

It is important to be watching for this change largely as a predictor of worsening afternoon weather which may affect your flying later in the day.

Avoid flight above building cumulus clouds

The temptation to quickly climb over even a thin layer of building cumulus should be avoided. This is a bad idea for anyone flying a low-performance aircraft. The rate at which these clouds build and their tops grow can easily exceed your climb rate, and they could quickly grow to an altitude which exceeds your aircraft's service ceiling.

In summer over the Colorado Rockies, the tops of afternoon thunderstorms can occasionally reach to over 30,000 ft. Even by midmorning, cloud tops well over 25,000 ft are normal. While these sorts of top elevations generally represent a few big thunderstorms rising well above the rest of the clouds, they can blend together until you are stuck trying to fly through a developing thunderstorm.

Returning to Telluride from Idaho in 2000, far ahead of us we noted a thin band of cumulus clouds over the Uinta Mountains. Since we could see over the clouds to terrain beyond, it appeared that the tops were slightly lower than our 11,500-ft cruising altitude. Nevertheless, since they were building, we began a slow climb anyway. As we drew closer, the clouds continued to build. Soon we were at 15,000 ft at our best rate of climb. By the time we reached 18,000 ft, the clouds were well above us and building even more rapidly. We were forced to circle down and go underneath. Total time suggests that cloud tops were rising at about 1000 ft/min, which would not be at all unusual. Most glider pilots have encountered stronger lift than that while circling below such clouds. Most mountain pilots have had many similar experiences. There is nothing unique about this one. Neither would it be unique if we had been forced to turn back and wait for better weather the next day.

On summary, flying on top of the clouds is a very acceptable procedure provided various conditions are met:
- You are flying toward better weather conditions.
- The weather is predicted to be improving.
- You have an easy escape route at hand.
- The cloud tops are not rising.
- You are in an aircraft capable of altitudes much higher than the cloud tops.
- You and the aircraft are prepared to go IFR if things change.

Flight under the clouds

If flying above the clouds has its hazards or is simply too high for your aircraft, should you fly below?

Certainly in the summer that is exactly what you will usually do in any small aircraft. Cloud bases over Colorado in the summer are often higher than 16,000 ft, giving you nearly 2000 ft of clearance over the higher peaks and more than that over major mountain passes. Cloud bases can descend, sometimes surprisingly rapidly when it starts to

Fletcher Anderson photo.

rain or snow, but mountain peaks are usually obscured long before mountain valleys, leaving an escape route open if you planned well. You can—*in fact, you must*—plan a route that allows you to land at an alternate airport if the cloud bases get lower or your flight is blocked by rain, snow, or lightning. But if you don't press your luck too far, there are enough mountain airports in major valleys to make that very possible.

Ducking through the passes under the clouds is a better idea in summer than in winter for another reason as well. In rain or mist, you always have at least *some* visibility, even if not enough for safe flying. In snow, visibility can actually go right down to zero.

The air above a cloud layer might be smooth, except perhaps as you flew through the very top of the inversion layer, or less likely if you flew into air above with a high wind gradient. The air below the clouds is probably bumpy because of the convective activity forming the clouds. It is worth trying a few different altitudes to find the smoothest level. Thermals building the clouds lose energy as they reach cloud base, so early on in the period of cloud formation getting up close to cloud base might work. Later on when the clouds are more developed, they produce strong internal updrafts of their own and still later overdevelop and produce strong downdrafts. At that point splitting the difference between cloud bases and the ground would be more comfortable.

*Typical summer weather over the San Juan Mountains of Colorado.
Notice that despite the considerable cloud buildup, cloud bases are
far above almost all these 14,000-ft mountains.* Fletcher Anderson photo.

Cloud suck

Paraglider pilots, whose maximum speed is less than 30 knots,
have a particular phobia about flying under large cumulus clouds.
The lift going up into the base of the cloud can exceed both their
ability to descend and their forward speed to get away. They refer
to this hazard as *cloud suck*. Most of the time in a small plane this
is an annoying problem you might have to deal with, but not a
lethal hazard. In the giant monster cells that romp across the Great
Plains in summer, it can be. In the 1930s, the U.S. Army lost its
largest dirigible to cloud suck.

Mammata

Once a cumulus cloud reaches a certain stage in its development, it
will rapidly overdevelop. The next stage will be rain, strong vertical
downdrafts, possibly seen as virga, and possibly microbursts. At that
point, you don't want to be under the cloud! A sign that this is about
to happen is *mammata* under the cloud, which was discussed in
Chapter 3. You definitely do not want to fly a small plane under a
large developing thunderstorm or an overdeveloped one which is
starting to downpour.

Why pilots find themselves in trouble beneath the clouds

Many, if not most, small mountain airports do not have instrument approaches or departures. The cloud cover in the mountains can frequently be obscuring the mountains themselves and roofing over the valleys. Yet on these same days flight within the valleys below ceilings of 2000 ft or more can be safe and reasonable. A flight from Grand Junction to Rifle, Colorado, up the valley of the Colorado River beneath a 2000-ft ceiling is very possible, with potential emergency landings along the interstate highway. From there flight onward to Glenwood Springs is a short enough distance to fly safely if you keep your eyes open and are prepared to turn back if the weather deteriorates. Aspen is another 40 mi up the Roaring Fork River Valley, with two private usable dirt fields in between. You could arrive (and I have many times) in Aspen in VFR conditions under a very solid 2000-ft ceiling, while overhead the big jets were all in dense clouds missing the approach because the decision altitude is 2500 ft. Of course this method would not let you continue any farther—Independence Pass and Haggarman Pass would both be completely obscured so you could never continue on to Leadville and the eastern slope.

Or could you? Perhaps a window over the pass would let you into the Arkansas Valley with three closeby airports, and leading to much lower terrain to the east? Windows over mountain passes are more likely than most people suspect. Wind is funneled through the pass, sometimes keeping the pass itself clear when the mountains on both sides are obscured. Referring to Chapter 4 on canyon flying might suggest the method of doing this, but you begin to reach the point where you are getting way out there beyond the nearest safe landing spot, and the rocks and clouds are getting very close together. It is at the "or could you?" phase when you can very suddenly see the situation change enough to get you in real trouble. This is just one more of the thousands of situations in the mountains when you have to force yourself not to do something you are pretty sure you could get away with. Things could begin to go very wrong. Consider just one of thousands of examples:

Dick Arnold recalls a Cessna 172 being delivered to Aspen for one of his students. The delivery pilot had some crop dusting experience and so was inclined to go low anyway. When the ceiling got lower, he pressed on up the valley. Now, in good weather, it is extremely obvious which valley leads to Aspen. This day in low visibility the pilot continued up what he thought was the Roaring Fork River Valley. In fact, he had failed to notice the confluence of the Roaring Fork River and Crystal River and was headed up the wrong valley. He continued on until after traveling about the right period of time and right distance, there was no airport and the valley was extremely narrow. In a very short time, there was only just enough visibility to perform a deliberate precautionary crash landing.

One of the many lessons here is that familiar landmarks look very different when a low ceiling obscures them. Underneath a ceiling is a perfectly normal method of flying. It turns into something else entirely when the cloud bases get lower and rain or snow begins to fall. The beginning of light precipitation often brings down cloud base with it. Once that happens, you are scud-running, which is a cause of many accidents every year both in the mountains and elsewhere.

So first let us look at how not to find yourself scud-running before looking at those situations in which you might elect to try it and how to save yourself when you do.

Scud-running

The hazards of flying beneath the clouds can be summarized as putting yourself in circumstances where the clouds get lower until you cannot maintain terrain clearance without getting into clouds; or rain or snow beneath the clouds reduces visibility until the same thing happens. If things reach this point in the mountains, you are in very serious trouble, indeed, because you undoubtedly stayed out of the clouds as long as possible, so now you are not only in clouds, but deep in the bottom of a valley surrounded by high mountains you cannot see! This danger is so readily apparent to most pilots that everyone tends to land well before things get quite that bad. Further, because mountain airports are down on the valley floors and clouds first tend to obscure the higher mountains and passes, it is usually apparent that you cannot possibly reach your destination long before your route back to the closest airport is socked in.

Scud-running is self-evidently a dumb idea, yet there are times when you might find it tempting to try. Before you do, ask yourself one last time how important it really is to get where you are going right now instead of tomorrow morning.

Conditions for scud-running

Before you consider scud-running, consider whether these conditions exist:

- *A flight toward weather known or predicted to be improving is a calculated gamble you might risk,* while pressing on into weather known to be deteriorating, hoping to arrive before it gets too bad, which is inviting disaster.

- *Flying when the weather is getting worse, hoping to get to your destination, means that if any of the inevitable minor delays occur, things become much worse and all your options disappear.*

- *A flight down a valley is typically a flight toward low terrain, more room to maneuver, and a higher ceiling.* A flight down the Roaring Fork River from Aspen to Glenwood Springs, Colorado, 2000 ft lower, is typically a safer bet than a flight up the valley in the other direction.

- *A flight toward lower terrain is typically also a flight toward higher ceilings.* Not only is there more room between the clouds and the ground, but also the cloud base as measured in feet above sea level can be higher. Clouds are more likely to be a scattered or broken layer than solid overcast.

- *A flight out of the mountains is typically, but not always, a flight toward better weather.* At least it is a flight toward fewer things to run into. After discussing conditions with someone who just landed, you might consider a flight west out of Telluride, Colorado, in very marginal conditions based on the knowledge that scarcely 5 mi west, not only is the terrain lower, but also you are out of the mountains proper and might well be flying under clear or clearing skies. You could sit at the end of the runway ready to go, just waiting for a window to open up.

- *A flight over terrain you are very familiar with toward a familiar destination is vastly safer in these conditions than a flight over unfamiliar terrain.* Beware, though: With the tops of all the nearby hills covered in clouds, it is very easy to misidentify landmarks.

- *A flight between two airports not very far apart makes it more likely you will reach your destination before the weather deteriorates.* If you guessed wrong and visibility deteriorated rapidly, you would need less time to return to your departure point. By extension, a route connecting several closely spaced airports gives you many more opportunities to stop and land if things get worse. A long flight between distant airports is asking for trouble.

- *A flight following a major highway is much better than a flight over open country.* Navigation will be easier because you will just be following a (hopefully) familiar road. The road itself is a possible emergency landing spot. Scud-running up the interstate highway over Kansas seems to be a very standard practice. If there is too much traffic on the interstate to land, you can use the frontage roads. Since land is cheap and the terrain is flat, it is easy to build an airport, and most are close to the road. Windmills are only about 30 ft high, so if you fly at 40 ft agl, you can miss them. Radio towers are the most common obstacle and are marked on the charts. Does flying around this low sound so ridiculous that you refuse to believe anyone would really try it? At least over western Kansas they do it all the time.

Beware of artificial obstructions!

I am aware of at least seven crashes that occurred in scud-running conditions in Alaska and Kansas when low-flying aircraft collided with the guy wires supporting radio towers. Airports are usually located near towns. Tall radio broadcasting towers are, too, and they are sited without regard to their possible conflicts with VFR aviation traffic. Despite the fact that they are supposed to be lighted, they can be very hard to see and the guy wires can be invisible. Several times under a low ceiling, I have spotted a tower nearly 600 ft high on the direct route from Greeley, Colorado, to Jeffco, Colorado, which is located very close to two intermediate airports.

In reading over drafts of this book, people told me of no less than five crashes in which their friends hit power lines in these conditions. No doubt many more than that have occurred. Power lines tend to follow roads, and they cross valleys rather high off the ground. In a word, power lines are very likely to be found right where you are likely to be when scud-running, and they can be absolutely impossible to see.

Single-engine charter pilots seem to be at risk for this accident. Single-engine charter, by regulation, must be VFR; so when the weather gets bad, single-engine charter pilots tend to scud-run rather than simply climb up into the clouds, file IFR, and fly an instrument approach to their destination. Are we looking at excessive machismo here? Not exactly. Very good, very careful, very experienced pilots do get caught occasionally. Once conditions deteriorate to the point of danger, they are too low down in the valley to go on instruments. The safest thing they can do is try to save themselves by scud-running.

How to scud-run

Again, the best advice is, Don't do it. Yet most experienced mountain pilots will probably agree that at least in a small low-performance airplane, cautious scud-running is probably less dangerous than actual IFR in the mountains. If you do elect to scud-run, or rather *if you find that rapidly changing weather has forced you into some scud-running* you wish you were not doing, consider the following advice:

- *Preserve the ability to turn back.* As in good visibility, it is crucial that you be able to turn around. This typically means flying up the side of a valley or canyon in order to have enough room to be able to make a nice, easy controlled turn back toward where you came from.

- *Keep your back door open.* Given the way clouds are constantly forming as air is lifted over the mountains, in low-visibility conditions, it is all too likely that rain or snow may creep up behind you and block your return to the last airport you flew over. Constantly checking behind you is necessary so that you can turn back as soon as something like that seems to be happening.

- *Maintain ground reference.* With the ground always in sight, you are preserving at least a degree of control over the situation. Once ground reference is lost, it could well be regained just as you fly into a cliff right in front of you. Because of the rugged terrain, once you lose ground reference, the first thing you are likely to find is a mountainside. Over flat country, you could settle down slowly back out of the mist. In the mountains you can't. Neither could you just change your mind and climb for safety. In a low-performance or any other aircraft, there would be little hope of outclimbing the surrounding mountains. A circling climb rather than a straight-ahead climb might be better, depending on your best guess about surrounding terrain, but your circle could easily drift into a mountain as well.

- *Fly very slowly.* Slower speeds mean that you will have more time to react if something looms up in front of you, and at slower speeds, your turn radius is much tighter. If you are flying very slowly with flaps down, you are essentially in a landing configuration if a promising landing site should appear right in front of you. The corollary of this is that certain types of aircraft lend themselves to safer scud-running better than others. A Beech King Air slowed down to 250 knots is going to slam right into a mountain before you even realized it was there. A Piper Super Cub slowed all the way down to 45 or 50 knots will get you into a tight situation much more slowly and require much less room to effect an escape. If (when?) things went completely to hell, the three times slower landing speed of the Cub would mean there might be a potential emergency landing spot the King Air could not use.

- *Fly very low!* This is in boldface italic because it is contrary to most other advice you will read or hear elsewhere. Most people will advise you to fly as high as possible in order to avoid hitting things. In the mountains, the things you might hit are very high indeed. Your best hope lies in seeing them as soon as possible so that you can turn around. My advice for these conditions is to try to stay 1000 ft below the clouds. Once things deteriorate to the point where that is impossible, mentally divide the space between cloud base and the ground in half and fly deep in the lower half. Flying up high near cloud base means it is very likely that you will be slightly in the clouds, or at least in haze, and forward visibility will be greatly reduced. The chances of inadvertently going right into the clouds are increased. Lower-lying scattered clouds may not be visible ahead until you are in them. *Flying much lower puts you in clearer air and lets you see much farther ahead!* You stand a far better chance of seeing which direction to go to get out from under the clouds. It is easier to tell if worse weather is creeping up behind you. Your view of the surrounding terrain is improved enough to greatly facilitate turning back. Your escape route can be much easier to spot as a patch of bright sun far away under the edge of the clouds. You are in a better position to detect whether conditions way up or down the valley look better or worse, and to plan accordingly.

- *Set your personal weather minimums very high.* In class E airspace, 3-mi visibility is the legal minimum. In class G airspace, which in most remote mountain areas is what you are flying in, you only need to be clear of clouds and have 1-mi visibility. These regulations were probably established by the Wright brothers in 1903 when the world was entirely flat and the only aircraft in the sky flew at about 14 knots. The most basic certified small aircraft flying today can eat up that 1 mi of visibility in only 30 s. Perhaps you could fly up the interstate highway over western Kansas in these conditions with some degree of safety. Over the mountains it would be perfectly legal but suicidally dangerous. Very few mountain pilots would take off for a mountain flight with ceilings of less than 2000 ft and visibility of less than 5 mi— and most of those would tend to be very young and invulnerable rather than exceptionally skilled or experienced.

Stand on the ramp at any mountain airport on a day with light snow or rain falling, under a 3000-ft ceiling with 10-mi visibility. Make an educated guess about the possibility of your reaching your destination without encountering weather just marginal enough to sock in one crucial mountain pass or ridge in your way. Can you even see some of the nearby mountains? Consider how very quickly falling rain or snow could make the conditions infinitely worse, and how far you might have to go to reach safety if it did. A lot of very experienced instrument-rated professional mountain pilots simply don't mess with these conditions in small aircraft.

Saving yourself when scud-running traps you

There is often no really good way to save yourself once visibility drops below a certain level, and that is why so many fatal accidents result. Obviously the best thing to do would have been to just land first. But you didn't do that this time, did you?

Fly toward low terrain

Whether you should employ this method depends very much on where you are and how well you know the surrounding area, but flying downhill down the bottom of the valley or canyon out of the mountains might save you.

Fly in low circles

Once you reach the point of having to try this, things have gone way past where they ever should have been allowed to. You are in very serious trouble, indeed. It is by no means likely, much less certain, that this or anything else you try can save you now. But here is what you might try:

Just before visibility goes right to zero, if there is nowhere possible to land, find somewhere you can circle with very good landmarks defining it. Slow way, way down, and circle around and around and around over your landmarks until hopefully visibility improves enough for you to escape. Obviously you have run out of options once you are caught having to do this. Your survival may very well depend on how soon the weather gets better, and of course it could also get worse. It may ultimately come down to whether you hit something solid or crash-land somewhere flat. If that is going to be the case, then choosing a spot to circle with a survivable spot where you could deliberately crash-land could be very significant.

Climb

Over flat country this might be the first thing to try. In the mountains you run the very obvious risk of climbing into something solid. Head away from known nearby high terrain up into the clouds. Collect your thoughts. Watch the attitude indicator and the other instruments. Call on the radio for help. Don't know a good frequency? *Try the local airport frequency first!* Most of the other aircraft in the vicinity use it. Try Flight Watch on 122.0, and they will give you a frequency for Center en route traffic controllers.

Every textbook I have ever read advises pilots to call for help on the emergency frequency, 121.50 in any emergency. Don't waste your time. Do you normally monitor that frequency? Neither does anyone else. After the tragic events of September 11, 2001, the FAA began advising pilots to monitor 121.50 all the time. Few people do it—they need the radio for other purposes. Aircraft in the air are advised by en route traffic controllers to monitor that frequency only after there has been an accident reported or an aircraft is reported overdue. Then what they are listening for is an ELT signal which broadcasts on that frequency. Only then will there be anyone to talk to on 121.50.

Unfortunately, in the mountains it is all too likely that there is very high terrain very closeby in all directions. You won't be able to talk to anyone until you are above it. Your best bet might be to circle up in a 20° bank turn until you know you are above everything. Again, this is not an inherently safe procedure. Circling straight up is intended to keep you clear of nearby mountains, but there is no guarantee you won't drift toward them. The climb has to be high enough to clear everything, which might mean a climb to at least 15,000 ft msl in icing conditions followed by a flight of an hour or more until you believe you are no longer over the mountains.

Precautionary landings

In his provocatively titled and extremely valuable pamphlet "How to Crash an Airplane," veteran pilot Mick Wilson notes that 18 percent of general aviation accidents result in fatalities. In Baker and Lamb's 1989 study of crashes in the Aspen, Colorado, area, 32 percent of all crashes in the mountains were fatal. Yet not all crashes are by any means the same. *Controlled flight into terrain,* which generally means you flew into the clouds and then ran into something, is nearly always fatal in the mountains. *Forced landings,* meaning, for example, that the engine quit and you are coming down, had only a 9.8 percent fatality rate. *Precautionary landings,* meaning that the pilots still retained enough control over the situation to decide to land and select the best place to do it, result in fatalities only 0.6 percent of the time!

Which one of these do you want to try, 100 percent fatality or 0.6 percent fatality? Admitting you made a mistake and getting on the ground anywhere is better than letting things go too long. Your responsibility is the lives of your passengers. The insurance company can take care of the plane. It may be psychologically helpful at this moment to remember that no airplane is worth more than $5000. That is, no aircraft policy has more than a $5000 deductible.

The lesson from these accident data suggests that you are at the very least 60 times more likely to survive *if you make the decision early* to find the best possible place to land and then do so before you lose control of the situation. Indeed, there is a 99.4 percent chance you will survive if that is what you do. Consider the following example:

At the time this incident occurred, Clem Kopf was an 1100-h private pilot and owner of a Cessna 210, which was down for maintenance. He had rented a Cessna 172 RG from Glenwood Aviation in the early spring of 1997 for a trip to Nevada. The weather on his return was not terribly bad over Grand Junction, Colorado, with numerous scattered snow showers in the vicinity, but plenty of room to maneuver around them. He elected to fly up the Colorado River Valley to Glenwood by way of Rifle, Colorado, rather than fly directly home. This turned out to be an exceptionally prudent decision. A gradually lowering ceiling began to roof over the valley. Scarcely 10 mi out from the Rifle Airport, his path was blocked by a very wet late spring snowstorm. Turning around to head back for Grand Junction, he discovered that a second storm had followed him up the valley. He was now trapped between two walls of very dense snow that was rapidly closing in on him. After soundly chastising himself for thinking he had to get back home in marginal weather, Clem quickly scouted the area for the best possible place to land. That proved to be the frontage road of the interstate highway (which was why he was flying that route to start with). He had time to make two low passes to check things out carefully and then landed on the third pass. People in nearby houses phoned the sheriff, who arrived just as the snowstorm hit. Clem explained what he had done and why. When the snow passed, Clem phoned ahead to confirm that his destination airport was now in the clear. (Cell phones are a wonderful piece of safety equipment.) The sheriff stopped traffic for him to take off, and he was landing at home 20 min later.

A spring day like the one which trapped Clem Kopf. Shelby Evans photo.

Because he made the decision to land early rather than fly around until things got really bad, Clem suffered nothing worse than wounded pride. Landing on an unobstructed road is just like landing on a runway, only longer. You don't get in trouble with the sheriff in this kind of situation, and even if you did, you would be in trouble on the ground that you might talk your way out of. Anyway, would you rather have your license threatened or your passengers' lives?

Roads are not perfect runways for two reasons: Utility lines usually parallel roads and often cross them. Traffic signs are often close enough to the roadside to snag a wing tip. In an emergency, snagging a road sign might damage the plane, but that isn't your big worry just then. Some utility lines are thin enough to just bust through, while others are strong enough to hang the plane from. If there is time enough to make a low pass and check for them, great. If not, you take your chances.

Don't you wish now that you hadn't tried scud-running? All your passengers are thinking that!

Walking out

This is as good a place as any to say that the conventional advice for what to do after a forced or a precautionary landing is to stay with the plane and wait for help. All planes these days are supposed to have an emergency locator transmitter. You should find that and make sure it is turned on. Make yourself as comfortable as possible. Start a fire both to keep warm and to help rescuers find you. Wait for help right where you are, instead of wandering off who knows where. Once searchers arrive in the area, they will be looking for a downed aircraft and will see smoke from the fire. They are not looking for, and easily will miss, someone walking out on the ground.

However, do not expect rescuers to be on the scene anytime soon. Once you are reported overdue, FAA personnel will begin a search by telephone. They will call all the airports you could have reached with your known fuel supply. They will call your home. Finally, having exhausted these possibilities, local law enforcement and/or the Civil Air Patrol will organize a search. This will take time to get started, and you may not be discovered for a day or two, if at all.

In Alaska, walking out could take weeks, even assuming you were walking the right direction. In Colorado, Wyoming, and New Mexico, hikers have been lost and wandering around for at least that long. They tell tales of climbing to higher ground to try to spot known landmarks, heading in what they believed was the direction of the nearest town, following the compass or looking for moss on the north side of a tree, and ultimately sometimes even stumbling back upon their own tracks.

The infamous Ken Torp/Rob Dubben party of cross-country skiers spent about a week wandering in all directions in the mountains near Aspen, Colorado, in the spring of 1995. Eventually an army (actually, the National Guard) of rescuers discovered the widely separated members of the party. Some had wandered here and there almost 50 mi the wrong way into Taylor Park.

When we were in high school, one of the graduation requirements was to climb 13,000-ft-high Mount Sopris. My stepbrother Steve was caught out in a blinding blizzard right on the summit with less than 10 ft of visibility. With no sun to guide him and no landmarks, he was soon hopelessly disoriented. He rescued himself in about 4 h. How did he do that? Why didn't he end up like Ken Torp?

My father taught us all, when we were very young, that anywhere in Colorado (or most other Rocky Mountain states), you will come to a road in less than a day if you walk only downhill. Also in less than a day, that same road followed downhill will get you to a house or other outpost of civilization. Steve just walked straight downhill to the highway.

In the winter, that day could stretch out into two or three, particularly if you had no skis or snowshoes. In Alaska, this won't work. After hiking for a month, you could come to a completely deserted stretch of beach somewhere on the Arctic Ocean.

Bad weather or deep snow may make it most advisable to just stay put. You have the aircraft for a shelter. It is easier to spot from the air. It has an ELT to attract attention. Richard Arnold recalls a couple of occasions when he ran Mountain Rescue in Aspen that people survived the crash, but got into serious trouble attempting to walk out. One was killed in an avalanche. But at least in Colorado, California, New Mexico, Wyoming, or many other places, if the snow is not too deep, you can walk to civilization in a day or less just by walking straight downhill. You don't need to know precisely where you are or what your intended destination is. In fact, you don't need to have a clue where you are, where north is, or anything else. All you have to know is what direction is down.

Fly around inclement weather

Since IFR is inherently dangerous, VFR on top has its own hazards, and VFR under the clouds sometimes leads to the possibility of scud-running, what is the very best thing to do? You could just fly around bad weather.

This seemingly self-evident option is too frequently ignored. Yet in the overall scope of things, the extra time required to fly around some weather to your destination is not really all that much (and it is still faster than your car). This can be a particularly useful and convenient option in the mountains because of the way the worst weather can collect over the high mountains, while major valleys and areas outside the mountains remain acceptably flyable.

When I was based in Glenwood Springs, a surprising number of our flight students and aircraft renters realized an aircraft was the easy way over the mountains. The aircraft could deadhead one leg; they could fly the other. I made numerous summer flights picking up bicyclists in Crested Butte, Colorado, and returning with them and the bikes to the grass runway in Marble, Colorado—a direct flight over Schofield Pass of 35 nautical mi. In winter I flew with backcountry skiers from Aspen to Crested Butte or vice versa, a distance of 45 nautical mi via Pearl Pass. Yet there were plenty of times when the direct route over the West Elk Mountains was completely obscured by weather. But recall that the worst weather usually forms right over the highest peaks. In those cases I sometimes flew the valleys south out of Crested Butte nearly to Gunnison, northwest to Kebler Pass, then north over McClure Pass to Carbondale, and finally southeast to Aspen—a distance of something like 190 nautical mi. Call that a flight of over 2 h, and it still beats a 7-h car trip. For that matter, at least once I found myself flying south to Gunnison, west to Montrose, north to Grand Junction, and finally east up the Colorado and Roaring Fork river valleys back to Aspen—a distance of well over 350 nautical mi. The 3.5-h flight still beats the car trip, and the extra 2 h in the air in the long run is a better idea than risking your life trapped in zero visibility in a narrow mountain valley.

In the spring of 1999 returning from Denver to Telluride in a Cessna 172, I encountered a wall of freezing rain just short of the Salida,

Colorado, airport. I turned back to the north, called Flight Service, and amended the flight plan to Eagle. After some of the weather passed over us, I took off again and made it to within 5 mi of Paonia, but had to turn back north to Rifle and wait again. Finally I was able to cross the Grand Mesa and return west of the Uncompahgre Plateau. All counted up, the less than 2-h flight home that began at 7:00 a.m. did not end until about 7:00 p.m. That is in fact slightly slower than traveling by car, but surely you recall the old adage: "Time to spare? Go by air!"

And you know, without my telling you, that the next morning, the weather was perfect.

Night flight over the mountains

Virtually every mountain flight instructor and every textbook ever published will tell you emphatically not to consider night flight in the mountains in a small plane. I am going to agree completely, but then I am going to backtrack a little.

Why not at night? A forced landing at night anywhere is substantially more dangerous than in the daytime because you cannot see well enough to select a suitable place to touch down. This is even truer in the mountains because there are so few satisfactory places for an emergency landing anyway. In general at night in the mountains, you should be flying major valleys where the lights of cars will show you the highway. The road is normally your best bet for a forced landing.

My advice for a night forced landing: Turn on the landing lights. If you don't like what you see, turn them off. You're going to land there anyway.

Terrain avoidance is far more difficult at night. Although you can usually see the mountains even on a night with no moon, you cannot make out very much or any detail. This in turn means you have no depth perception, so you can't tell if a mountain is 0.5 or 10 mi away.

Inadvertent flight into clouds at night is always a risk because you can't see the dark clouds against the dark background. Should that occur, you are at greater risk in the mountains because clouds in the mountains so often cloak the mountain peaks themselves.

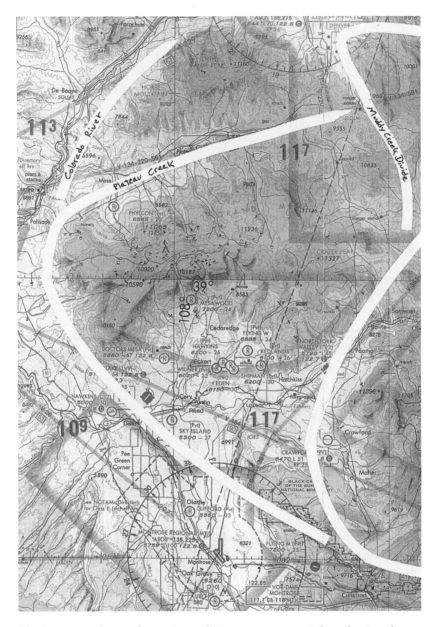

The long way home from Crested Butte to Aspen, Colorado. See the explanation in the text.

High-altitude winds that produce mountain wave or rotor do not necessarily go away at night. Strong sink could force you below your intended altitude just as it does in the day, but at night you would have difficulty telling where to turn to escape.

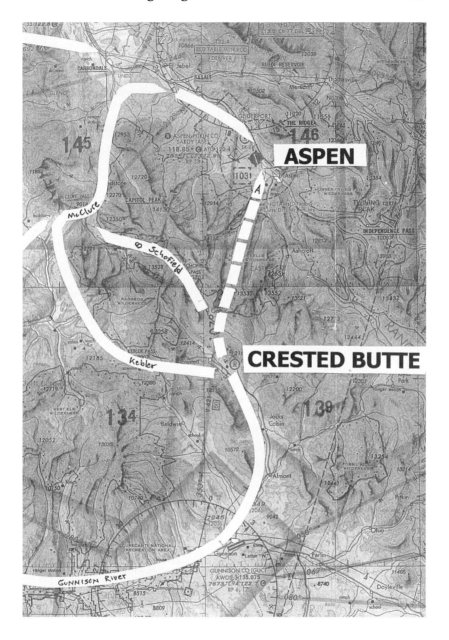

So conventional wisdom has it that night flight in the mountains is a bad idea, and it is fairly obvious why. But you might want to consider the idea anyway. Here is why and how.

Why at night? Night weather is often better than day weather, particularly in the summer. Early morning weather is normally better

Night flight. Look more closely at this photograph of a mountain wave lenticular cloud. The tiny spots of light on the mountain to the right are the headlights of snowcats grooming the ski runs. The white spots in the sky are stars. Commercial glider pilot Vincent Esposito took this photograph over Bear Creek, Colorado, at 2:00 a.m. on his way home from his night job as chef of Campagna restaurant in Telluride, Colorado. The cloud is being illuminated by the light of the full moon. Mountain weather does not go away just because it gets dark!
Vincent Esposito photo.

than weather later in the day at any time of year. A night departure can mean a daylight arrival at an optimal time.

Convective turbulence is the direct result of heating from the sun. After the sun has set, the turbulence goes away. Turbulence caused by strong winds over rugged terrain can continue on through the night, but nighttime temperature inversions often mean that there will be much more stable air beneath the inversion layer. Particularly in summer, less so in winter, but often enough, even when the day has seen unacceptable low-level turbulence, the night could very well be completely smooth and the surface wind could actually be calm.

Your circumstances might make it most prudent to arrive at your destination during the early morning hours before daily wind and turbulence begin. This brings to mind the backcountry landing strip at Sand Wash, Utah, where Slickrock Air Guides and Redtail Aviation regularly deliver river trip passengers. The strip is perched on a narrow bench above steep cliffs, well below the rim of the canyon proper. Once the summer midday or afternoon winds start blowing, it can be excessively dangerous to try to land. Rather than risk not being able to get in safely (or risk scaring their clients so badly that they never book again), the charter pilots elect to fly very early in the morning. Both those charter companies employ pilots with exceptional backcountry skills, but foremost among those skills is the ability to use good judgment.

If you (or when I) need to deliver a passenger there from a mountain airport 3 h or more away, you might very well decide that the safest thing to do would be to depart before dawn in order to arrive by 9:00 a.m. or earlier.

When you are leaving Aspen, Telluride, Glenwood Springs, or other Colorado mountain airports for Denver on a summer afternoon, the prospect of flying out through major afternoon thunderstorms might not seem terribly intelligent. You know that the storms will dissipate after sunset. You could then depart in the mountains in the early evening and arrive in Denver at night.

In both previous examples, you are departing a mountain airport in the dark, headed for a destination out of the mountains. You could also depart in the dark while expecting to arrive at a mountain destination after dawn. A few times, I have had passenger pickups in Aspen scheduled to happen the moment the airport opened in the morning. I would arrive 10 min early and circle over Aspen Mountain until the tower came on frequency.

Flights from one mountain airport to another at night in a small plane make less sense, and you should really be asking yourself why the flight is necessary.

How to fly the mountains at night in a small plane. En route, the basic technique is quite simple: You just stay well above the nearest high terrain. This means that on a flight over the middle of the Rocky Mountains, you will be at least at 15,000 ft msl, and would prefer to be 1000 ft higher still. We are getting into altitudes which require oxygen and altitudes that not all small planes can reach. This is another of those strategies that not all pilots in all aircraft can use.

In the winter, snow-covered mountains are nearly as visible under a full moon as they are in daylight. On those nights, you could fly cautiously and maintain terrain separation by using the same practices you do in the day.

Some aspects of navigation at night are not difficult. Radio or GPS navigation is the same day or night. Identifying physical landmarks can be difficult or impossible on any night with little moonlight. On the other hand, the lights of a town or the headlights of cars on a major highway are easier to spot and identify from a distance at night, although of course it is also probably easier to identify the *wrong* town or highway!

Climbing out of a mountain valley, you have to decide between two strategies based on your local circumstances. You could plot out where the nearby high terrain is and select a route, using radio navigation aids, which avoids it. That is, you could depart tracking a VOR radial. Instrument departures follow similar procedures.

The other method is to circle up. You climb up in a circle, staying directly over the lights of a town. The higher you get, the wider the valley, so the farther from the rocks you are. This is very easy to do, but of course you wouldn't try it in a very narrow valley.

Departing the very narrow valley at Glenwood Springs, where there are no navigation aids and the valley is too narrow to circle, we would fly a known heading until we were above the rim. (Glenwood Springs has a peculiar city ordinance that prohibits night landings, but not night takeoffs. When the ordinance was written, noise was not the issue it is today, but a landing pilot had hit deer on the unlighted runway. The procedure in those days was to buzz your house to get a family member to drive out and shine car headlights down the runway so you could land.)

Descending into a narrow valley to land at night is something best done only in very good weather.

If you are going into an airport with an instrument approach, great! You simply fly the instrument approach procedure, which guarantees your terrain separation. If your airplane doesn't fly high enough to get to the altitudes where the procedure normally begins, you can still use the final segment. Not instrument-rated, or not flying a plane which can use all the available aids? You can still devise your own straight-line descent route into the valley, either tracking a VOR radial or flying a straight line to the airport on a predetermined heading using your GPS.

Visual descents are easy enough to do in concept, although cautions about doing so in marginal weather cannot be overstated. These descents rely on your being able to see lights on the ground. These might be the runway lights, but equally useful are the lights of the nearest town. Once the lights of your destination are clearly in sight, you begin a constant rate of descent directly toward the lights. Provided nothing obscures any of the lights, there is nothing between them and you that you could hit. Watch very closely. If one or two lights begin to blink out, you are descending too steeply and a mountain or ridge is rising up into your path. Add power and flatten out the descent.

It is far, far safer to try these night descents into airports you are already very familiar with in daytime. Knowing the location of the nearby high terrain is better than guessing about it after deducing it from the map. It is very easy to be fooled into thinking that a mountain silhouetted against the lights of a town is just a big empty field on the ground with no lights. If you already knew a mountain was there, you would correctly interpret what you were seeing. Aspen, Colorado, for years allowed night landings only in instrument-equipped aircraft for instrument-rated pilots *who had made three daytime landings there in the past 6 months*. This would be a good personal limitation for any pilot landing at any mountain airport at night.

Fly on another day

What a concept! Yet this exceedingly self-evident option is also too often ignored. The morning after my epic 12-h long struggle to get home from Denver, despite a prolonged monsoon season, skies were clear all over the mountains until about 10:00 a.m. and still reasonably flyable VFR until noon. It would have been very easy to spend a comfortable night in Denver and still make any scheduled commitment the next day. If the schedule were too tight, it would have been possible to rent a car or book a flight with the airlines.

All pilots, certainly including me, fall into the psychological trap of believing a flight has to be accomplished *right now* that could probably be delayed for days or put off entirely. Pushing on into deteriorating weather, hoping to reach home before it gets too much worse, is a leading cause of small aircraft accidents everywhere. In the mountains, you have fewer means of escape once the weather actually does get worse.

At every mountain airport where I have worked, I have watched numerous low-time private pilots depart into horrendously dangerous weather conditions predicted to get worse, while the locally based high-time professional pilots were staying on the ground. At Telluride alone, I have watched charter pilots Galen Rassmussen, Peter Lert, and Josh Thompson, all flying large twin turboprops, and airline senior captains Jeff Garcia and Troy Beattie, both flying airliners, blow off an approach to land in turbulence and divert. These are all people with several thousand hours of mountain experience and hundreds of Telluride landings flying very capable, large, familiar aircraft. Moments later, I have then seen low-time private pilots depart in their Cessna 210s or Beech Bonanzas in the same conditions and get severely thrashed.

The pilots departing get hammered, frequently get slammed right back down on the pavement, barely clear the fence, drop out of sight below the runway into the valley, get rolled past 60° bank, tossed upward thousands of feet—but, thank God, they almost always keep going west out of the valley. We see about one crash every 2 years at Telluride, usually involving this scenario. As this is being written, there is a brand-new Cessna 206 from out of town waiting for a new propeller after an attempted takeoff had it wheelbarrow sideways down the runway on the nosewheel and hit the pavement with the propeller.

Why do so many people do something this dumb? For that matter, why do all of us consider doing something this dumb?

They are busy people with schedules to meet who have to be back home to make a business meeting the next day. They are able to convince themselves that conditions are not quite as bad as they themselves could easily see they were, but for the demands of their schedules. In 1991 I was guilty of this obvious error, convincing myself the weather was different from what I knew it was, launching my paraglider into obviously dangerous conditions, and crashing just because I had a strong urge to go flying.

Part of this excessively risky behavior can be attributed directly to lack of mountain experience. More important than just having a schedule they need to meet, these are lower-time pilots, most of whose experience has been at low altitude in flat country. They have taken off from flatland airports in stronger winds many times without incident, and they assume the same will be true in the mountains.

The mountain-based professional pilots electing to stay on the ground know from previous experience that a takeoff would be right into severe low-level turbulence below the ridge tops, or possibly right over the runway, that they are not going to risk.

Planning to depart in questionable weather? Why not discuss your plan first with some of the local pilots? Their local knowledge may be telling them something your more general knowledge and experience is not detecting.

Planning not to go? This is never a bad decision, but experienced local pilots may know why conditions are not actually as bad as they look.

6

Almost home

Descents

Flatland pilots seldom give the descent to landing very much thought. A very straightforward calculation tells you to divide your altitude by 500 ft/min to get the time needed for the descent. Then multiply your speed on the descent by the number of minutes required to descend to get a distance out from the airport where you will begin to come down. Are you too lazy to calculate this (as I am)? Then you just start down when the airport comes in sight. A descent into a mountain airport is not quite that simple.

Because the mountains are very high, while mountain airports are located in the valley bottoms, the descents into most mountain airports have three things in common: The descent is very long and there is little room to do it, so the descent is very steep. Planning an approach to minimize these linked problems is discussed soon.

A very long, very steep descent can pose certain problems for your engine, as well as speed control problems which can cause trouble in turbulence. Let us first consider the engine.

Engine management

Aircraft engines are designed to be as lightweight as possible. Light weight is achieved in part by using the thinnest possible metal in the engine cases, cylinders, and cylinder heads. These parts expand when the engine is hot and contract when it cools.

The next time you visit an airport, look over a copy of *Trade-A-Plane* (P.O. Box 509, Crossville, TN 38557, www.trade-a-plane.com). In particular, look over the ads for Cessna Turbo 206s and Turbo

210s, or Beech Turbo Bonanzas. These aircraft are powered by a Continental TIO520 engine. Quite a few, if not the majority, of these ads will carry the phrase "XXX hours STOH," where STOH means "since top overhaul," or "XXX hours since six new cylinders." After only 500 h of flying, it was necessary to remove and overhaul or replace all six cylinders—at a minimum an $8000 repair! This is a very smooth, very capable engine, but it is an easy one to abuse. A long climb at high power settings can easily overheat the cylinders and the oil, resulting in premature wear, if not the actual destruction of the cylinder bore. This engine is also extremely sensitive to shock cooling, which results in cracks in the cylinder head, usually near the spark plug holes or the fuel injector nozzle. Managed carefully, this engine lasts a long time, and the turbocharging makes it ideal for high altitudes. Abused by overheating and shock cooling, this engine can have a very short but expensive life.

Avoiding shock cooling

Shock cooling is the excessively rapid cooling of the engine. Very rapid cooling can cause cracking, particularly as cooling and contracting will not be uniform over the entire engine. It can be a cause of cracking of the cylinder heads as well as other engine parts. For any aircraft piston engine, you should reduce power gradually in steps to prevent this and possibly other problems. Here is an outline of how to do it, but as always in a case like this, consult the operator's manual to determine the method most precisely appropriate for your aircraft.

For a naturally aspirated engine with a fixed-pitch propeller. On the tachometer, there is a green arc for normal engine operating rpm's. Pull the power back to the bottom of the green arc, and fly at that power setting for at least 1 min. Then, *if necessary,* pull on the carburetor heat and reduce the power to less than 2000, but no less than 1700 rpm for the rest of the descent. It is my practice, when I am towing gliders or dropping parachutists with this type of engine, to go to full flaps before beginning the descent. The engine will have all but overheated on the long, full-power climb, while at the same time it will be necessary to get back on the ground as soon as possible to pick up the next load. By descending with full flaps, airspeed can be kept low with some power left on, so the flow of cooling air through the cowling is reduced and the engine cools more slowly and uniformly. If nothing else, pulling the power all the way back to idle will cause a prolonged episode of backfiring.

For a naturally aspirated engine with a constant-speed propeller and cowl flaps. Close the cowl flaps. Reduce the power to about the middle of the green arc on the manifold pressure gauge for 1 min. After 1 min, reduce the power to the bottom of the green arc. If necessary, use the flaps to control airspeed. Use propeller pitch control to keep the propeller turning in the slower part of the green arc on the tachometer (until you are on final approach and go to climb pitch in case you have to go around).

For a turbocharged engine with constant-speed propeller and cowl flaps. These engines require the most engine management on a descent, *and* they are the most expensive to repair. You are probably cruising at a manifold pressure higher than sea-level air pressure. That is how the engine develops so much power, which in turn is how you are able to cruise so fast. A basic rule of thumb for these engines is as follows: Close the cowl flaps, if they are not already closed for cruise. Reduce the throttle setting by 1 in of manifold pressure, and fly 1 min. Then reduce it another inch and fly another minute, and so on until the manifold pressure is reduced to the very bottom of the green arc. Adjust the mixture for smooth operation.

This is going to require some planning on your part because this form of engine management is a 5- to 10-min process which you will want to complete *before* you begin your final descent. If you start descending first, you will be unable to keep your speed within reasonable limits because you will be flying at too high a power setting.

Leave the propeller pitch alone at cruise rpm until you are on final to land. Control your speed and rate of descent as much as possible by lowering the flaps and landing gear rather than just by reducing power. (*Note:* This advice may be contradicted by engine-specific advice in the manual, which may suggest using higher rpm's with low manifold pressure to avoid detuning an engine with dynamic crank balancing weights, or may suggest lower rpm's to slow the flow of cold oil through the oil cooler. The manual always has the last word.)

Are you in a real hurry to get down? Then reduce manifold pressure 2 in and fly 2 min. Reduce it 2 in more, fly 2 min more, and repeat until you are down to 20 in of pressure. After that, you can throttle back pretty much as necessary. Pushing the constant-speed propeller to high rpm's will make it act as a brake for a descent at low power settings, although this will do nothing to extend the life of the engine. (This can be extremely effective with large turboprop engines, but less effective with small piston engines anyway. Don't even consider this with a geared engine—the results would ultimately be very expensive.)

Remember that these are just very basic guidelines. The operator's manual is the definitive set of instructions!

Avoiding overcooling

Overcooling a small aircraft engine on a descent outside of the mountains is seldom an issue. If you are flying at, say, 2000 ft msl and descending down to a landing at 500 ft msl, then you will only be cooling the engine for a total of about 3 min. You are unlikely to overcool it, whatever the temperature. If you are crossing Loveland Pass at 15,000 ft msl, descending to land at Centennial Airport at 5880 ft msl, then you could be subjecting your engine to very powerful cooling for 20 min.

While generally it is not an issue in summer, overcooling the engine can create problems during the winter months. During those months, the cooling air rushing into the cowling and over the engine and oil cooler can be at a temperature well below freezing, if not in fact well below zero. Shock cooling is more likely because of the very rapid cooling effect of air that cold. The result of a very prolonged descent with low power and air that cold cooling the engine is an engine cooled to well below the temperatures allowable for full-power operation. You can see this result by looking at the cylinder head temperature and oil temperature gauges. As you descend, the needles will begin a precipitous drop and will finally stabilize below the green arc on the gauges.

You would prudently allow the engine to warm up into the green arc on the temperature gauges before takeoff. Otherwise it would not reach full power, engine wear would be excessive, oil seals might be blown by high pressure, and possible cracking of the cylinders and heads could result from nonuniform heating and expansion. These same factors would come into play if you were forced to do a go-around instead of to land with a very cold engine. Further, you would not idle an engine at very high rpm's until it was warmed up. On a descent, an overcooled engine would reach very low temperature but would still be turning over at much higher than idle rpm's. The bearings, cylinder bores, piston rings, and oil seals would not respond well to this treatment.

The method of avoiding this trouble is to keep some power on through the descent. You could do this by using more than normal flaps and lower descent speeds. You could lower the landing gear early rather than reduce power. You could also simply take a little more time to make a longer descent at a lower descent rate.

Speed control

On a really smooth, calm day, it is not necessary to control your speed on a descent very much. You just point the nose down, let your speed climb up into the yellow arc on the airspeed indicator, and arrive that much sooner at your destination. The last day we had weather like that in the mountains, Richard Nixon was still in the Whitehouse.

Descending into any deep, narrow valley in the mountains, some degree of turbulence can almost be assumed. If the turbulence is bad enough, then you have to fly through it at less than your aircraft's maneuvering speed, or V_a. For a short descent, simply reducing power is an acceptable way to do this. For a very long descent, particularly in winter, your engine could end up being overcooled by the end of the descent. Lowering the flaps (and deploying the speed brakes and lowering the gear, if so equipped) and then starting down will give you an acceptably steep and rapid descent without building up excessive speed. Note a few things:

- Slow down, lower the flaps, and *then* descend. If you just drop the nose first, you may never slow down enough to lower the flaps, and by the time you do, the descent angle in all likelihood will have been nowhere near as steep.
- Maneuvering speed V_a is slower with a lighter load. Check the manual or calculate the actual value for your load. In actual turbulence, you will have to descend at slower than book maneuvering speed. You might prefer to do that for comfort purposes anyway.
- With the flaps down, you will have to fly well below V_a and in fact well below the top of the white arc on the airspeed indicator to avoid flap damage in turbulence. More than just flap damage could occur. Maneuvering speed V_a is a factor of the aircraft's stalling speed. It is the speed above which a

full abrupt deflection of the controls could cause airframe damage, and below which full abrupt control deflection would produce a stall of the control surfaces. Presumably you don't fly around making full, abrupt control deflections, but the same speed translates to your turbulence penetration speed. Lowering the flaps changes the airfoil of the wings. Now stall speed is lower, and therefore so is maneuvering speed. Damage to the entire wing could result from flying too fast in this configuration in turbulence. Some, but by no means all, aircraft operator's manuals list maximum *g* loads for different flap configurations. Some people suggest this means you should not use flaps in turbulence. No, you can use them. You just have to be vigilant about keeping your speed under control.

- If you are flying a complex aircraft, don't hesitate to use other configuration changes to increase drag for speed control. Descending from at least 15,000 ft up to around 18,000 ft over the Continental Divide on the way down to only 5000 ft in Denver, when flying a retractable-gear single-engine aircraft, we often drop the landing gear as we pass Mount Evans, some 30 mi out from the airport. Some high-performance aircraft have been retrofitted with speed brakes on top of the wing similar to miniature glider spoilers.

- *Power controls altitude, pitch controls airspeed.* You control your speed on a descent by adjusting your descent attitude. If you are going too fast, lift the nose slightly rather than just reduce power. If your rate of descent is too slow, reduce power or increase drag by lowering the flaps or gear rather than just dropping the nose.

Carburetor icing

When air expands, it cools. If temperature and humidity are at the right levels, the expanding air will cool enough to form condensation and the condensation may occur in the form of frost. When you pull back the throttle on an engine with a carburetor, you are turning a small butterfly valve in the carburetor. That forms a constriction in the engine's air intake. When you are descending (occasionally even when you are taxiing), the engine is sucking air in through that constriction. Downstream in the airflow from the butterfly valve, the air expands again. If the air temperature is between about 60°F and freezing and there is moisture present, the expanding air will cool and form frost in the carburetor. This frost accumulation is called *carburetor ice*. It can severely restrict the airflow into the engine, causing very noticeable loss of power. It can even cause the engine to stop running.

Although the right amount of moisture might not always be present in the mountains, the right temperature range to form carburetor ice often is. Fortunately for you, the engine manufacturers are aware of this problem and have fitted a carburetor heating device. This is simply an alternate source of engine intake air which draws hot air from a cuff fitted around one of the exhaust pipes. There is a knob labeled carburetor heat next to the throttle. Whenever you reduce power below cruise settings to descend, you pull on this knob. The operator's manual will tell you exactly at what rpm's it is necessary.

Note that while carburetor ice will not form if the temperature is much colder than or much warmer than a certain temperature range, and will not form unless there is sufficient moisture, it is good practice to simply use carburetor heat for *any* low-power descent rather than try to calculate whether it is needed. No harm is caused to the engine by its use for this purpose.

Descent planning

Finally, if terrain allows it, plan a descent route that does not require an unusually steep descent. While there is little you can do to reduce the total number of feet you have to descend, you can sometimes pick a route that allows you to start the descent early.

Pilots familiar with the area in small aircraft coming into Aspen, Colorado, from the south will cross one of several passes as low as possible and descend down the bottom of the very narrow Maroon or Castle Creek Valley. Then instead of circling down, they pop right around the corner over the Buttermilk ski area parking lot exactly on glide path to land on runway 33. This procedure has the very desirable side effect of really wowing the passengers!

Unfamiliar with the area? Then this might not be an option because you might have trouble identifying the right pass or locating the airport. Rather than dive into the valley, you could just plan on a long descent out to the west before flying back east to the airport.

Planning the approach. If he is going to have to go around, this pilot will have to make his decision early in order to turn to the right and aim toward the low terrain that way. Notice that like most experienced small aircraft mountain pilots, he is making a slightly steeper than normal approach. John Kounis photo.

Approaches

Avid readers of flight instruction manuals or the *Airman's Information Manual* know that the correct way to make an approach to an airport is to cruise along until the airport is in sight. Then you begin a descent at a constant 500 ft/min until you are at something fairly near pattern altitude. Then you drop down to pattern altitude and enter a standard traffic pattern on a 45° angle to the downwind leg.

This is a beautiful theory, but coming in to most mountain airports, it just can't be done. *The approach to a mountain airport is virtually always dictated by terrain.* Long, straight descents to join up with a standard traffic pattern are often impossible. Very long, winding approaches up the valley are far more likely.

Before you start a trip to an unfamiliar airport, get out the chart and consider your approach route. Mountain airports are virtually always situated on the floors of valleys. Often the valleys are narrow enough to make an approach in over one wall of the valley too steep to accomplish without circling. If you are lucky, you could be headed for an airport such as Salida, Buena Vista, or Leadville, Colorado, all spaced along the bottom of the broad Arkansas River Valley. The Arkansas Valley is more than wide enough to fly a standard traffic pattern. All the airports in it can be approached by flying either up or down the valley, so traffic pattern and runway choice are essentially dictated by prevailing wind.

Next in difficulty of approach in this context might be airports such as Aspen or Telluride which are located up at the head end of their respective valleys, meaning that for most practical purposes, all approaches will be flown up the valley, and all departures flown down the valley. *Opposite-direction traffic on the runway is the norm, not the exception, for any mountain airport.*

Now add the constraints of a narrower valley which restricts the traffic pattern. Aspen is next to a bluff adjacent to the west side of the runway, so the traffic pattern for runway 15 is the standard left pattern all the textbooks mandate, but the pattern for the opposite-direction runway 33 is a nonstandard right hand. Thus the two traffic patterns occupy the same space, but flow in opposite directions! At Telluride, both patterns are south of the runway; in Rifle, both are north; in Crawford, both are south; in Glenwood, both are west; etc. See a pattern developing? *Opposite-direction traffic in the traffic pattern is perfectly normal at any mountain airport.*

I suppose the next step beyond this is airports such as the private field in Marble, Colorado, or various airports along the Salmon River in Idaho which are essentially one way—you can virtually only land going up valley and only take off down the valley. No traffic pattern is possible *either* side of the runway. For most pilots in most airplanes, a field like this is also one-shot: once on final approach headed up the valley, you *will* land. The climb is too steep and the turn too sharp to go around.

Then we move into the realm of a remote dirt field such as Mineral Bottom, Utah, right on the bank of the Green River at the very bottom of a vertical-walled sandstone canyon as deep as it is wide. Landing and taking off are no problem in the right plane, but the route in and the route out are way down in the canyon bottom coming around the bend. The runway is not even in sight until you are on very short final. If you are forced to go around, you continue flying up or down the canyon around the next (very close) bend. For even more passenger thrills, I prefer to fly a base leg down an even

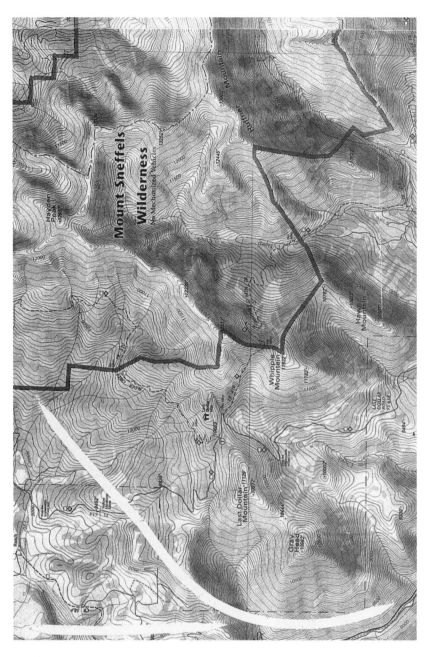

Traffic congestion at Telluride, Colorado. Despite the seemingly endless amount of airspace around the airport, all the approaching traffic is coming in from the west, and all the departing traffic is departing toward the west. Everyone coming into the valley from or departing to the north cuts around the same corner. Everyone coming from or departing to the south crosses the same narrow pass 6 mi away.

Finally, the traffic pattern is only on the south side of the runway, so right traffic for runway 9 and left traffic for runway 27 occupy the same airspace. Opposite-direction traffic using the same runway is the norm. Gliders in the area may be using a different frequency or have dead radio batteries and usually completely ignore the standard traffic pattern. Large jets are on a Denver center traffic control frequency rather than unicom.

Wilson Bar Airport on the Salmon River, Idaho. The narrow confines of this canyon restrict the approach path, midday turbulence is almost a given, and this is really a one-way-in, one-way-out strip except perhaps for a very short-takeoff aircraft like this Husky. John Kounis photo.

Mineral Bottom, Utah. Compare this with the photograph above. You have to fly down the narrow inner gorge of the Green River in Utah, and the runway is not in sight until you are 1 mi out. Sequence 1: Can't see it yet, only 2 mi out.

Sequence 2: Closer, and note the typical turbulence in a narrow canyon.

Sequence 3: Finally the runway, 0.5 mi out. To go around, you continue on down the canyon. The turn is not as sharp as it appears to be, and you sometimes find that campers have pitched tents on the runway. John Kounis photos.

narrower side canyon before turning final into the main canyon. Red Tail Aviation and Slickrock Air Guides in Moab, Utah, make a regular summer business of dropping river runners in places like this. Seemingly one-half the state of Alaska is just as difficult to approach, but nobody ever smoothed out the landing spots. For pilots experienced in this kind of flying, it is, if anything, less difficult for them than the wide open, unobstructed approach to the 3-mi-long, 200-ft-wide runways of Denver International Airport.

Experienced is the magic word. You save yourself considerable stress by making your first visit to a place like this with another pilot who regularly travels there.

Descending into a valley

Having decided that flying along until you see the airport and then descending are not always going to be practical, you will plan a route to first descend into the valley where the airport is located and then fly up (or down) the valley to the airport.

An obvious but usually neglected source of help here is the telephone. Call the airport. Ask whoever answers the phone how local

Not really a bush plane, the Cessna 206 is the workhorse for most single-engine charter operators worldwide. Observe how closely the pilot needs to fly to the rock wall to set up for a landing near the Green River in Utah to drop off passengers for a river trip. John Kounis photo.

pilots flying aircraft like yours come in to the airport. If the person answering the phone doesn't know, she or he can refer you to either a local instructor or a local-based pilot who can help you.

The other source of information is the chart, or perhaps if you have a copy, the local U.S. Geological Survey (USGS) topographic map. Spread it out on the floor and start studying. A standard descent for most small airplanes works out to about 1000 ft every 2.5 or 3 mi. Start scaling off distances and altitudes from the airport, and see where that puts you.

Approaching Telluride from the west? This one is a no-brainer. Just come straight in. Approaching Aspen from the south? Now there are some choices to be made. It looks as if it would work out just right to cross one of several passes over the West Elk Mountains as low as possible and descend down any of several steep, narrow canyons. In fact, this will work out just great, but most pilots don't always elect to do it that way. Why not?

Predicting areas of turbulence

You know either from previous flying experience or from reading the chapters covering the en route portion of the flight in this book

Grand Canyon West Airport in Arizona is a more extreme example of the kind of desert or mountain airport where midday wind or convective turbulence can limit small aircraft to the early morning hours of the day. The airport is situated this close to the rim in order to serve hikers going down into the canyon. John Kounis photo.

The proximity of numerous cliffs to the runway at Telluride produces turbulence with only moderate winds. On the approach to runway 9, shown here, pilots can keep out of some of the turbulence from south winds by flying the approach left of the runway centerline rather than directly over the narrow canyon. Shelby Evans photo.

that narrow mountain canyons are collectors of sinking air and turbulence. You also know that the lee side of a ridge is never the place to be flying. The dangers of descending narrow canyons or descending the lee side of a ridge are far, far less than the dangers of trying to climb in the same place, largely because you are already headed downhill toward lower terrain, so you are already on your escape route, not just set up for it. But why subject yourself to the discomfort, the stress, and possibly even the loss of control from turbulence worse than anticipated?

Call Flight Service to get the predicted winds aloft. If winds are strong, start planning again.

Unless you are very certain that you are flying into very, very calm conditions, it is best to put the temptation to fly down a very narrow, but very scenic and dramatic canyon out of your mind. Next mentally block off the leeward side of any closeby ridge or peak. Now go back two paragraphs and begin scaling off altitudes and distances again to plot a couple of likely choices. This could very well result in a route of choice that seems to be very long and way out of your way. To go back to the example of Aspen, most pilots in small aircraft coming up from the south will cross McClure Pass, 20 nautical mi west of Aspen, fly straight north 10 mi past Mount Sopris, and then fly 17 mi southeast back up the valley to Aspen. Depending on where you were arriving from, this might seem like a 25-mi or more detour, but it could avoid considerable turbulence, very close proximity to very rugged terrain over very high passes, and descents through very narrow canyons.

Descending through lift

Astute readers will note a preference for descending anywhere but on the lee side of a ridge. That is correct because while light turbulence is uncomfortable, moderate to severe turbulence is dangerous. Should you then descend down the center of the valley or down one side?

If the air is relatively smooth, you can descend anywhere you want. Right down the center of the valley will work fine since you are already headed toward low terrain and don't have to be planning to suddenly turn around. Many people will suggest that you habitually

Descending through lift. I don't know who this Husky pilot is, but he flies precisely the way this book advocates, so we see his picture a lot. On approach to Loon Creek Airport near the Salmon River in Idaho, he is flying way over against the side of the valley in smoother lifting air rather than through turbulence in the middle of the canyon, and he has left plenty of room to turn around if he has to. John Kounis photo.

keep to the right side of the valley to avoid other traffic, just as you would stay in the right lane on the highway. This seems like a reasonable idea, but there will be times when you can't do it.

When wind or turbulence is strong, safety will dictate staying away from the lee side of a ridge, regardless of your direction of flight. The upwind side of the valley will be filled with air tumbling down into the valley. You will have sink, which might get you down quicker, but also turbulence best avoided. The downwind side of the valley will have air compressing back together and lifting up the valley wall. *My preference is almost always to descend through this lifting air because it is smoother.* Even if it is not all that smooth, when you hit turbulence, would you rather it pushed you down toward the rocks or up away from them?

Planning an approach for conditions of reduced visibility

You have already noted that mountain peaks become obscured by clouds long before valleys do. For very obvious reasons, no one with any sense at all is going to suggest that you fly into a mountain airport for the first time in conditions of reduced visibility. You might be tempted to fly to a familiar airport in such conditions, however. If you do, your route will almost certainly be up the valley, and your point of entry into the valley may prove to be very far away. *By flying up the valley, you are probably flying toward worse weather!* This is a time to set your personal weather minima much higher than you otherwise might. It is critically important in these conditions to be sure you are not running a risk of being trapped by deteriorating weather. You must have a wide-open escape back out the valley to an airport known to be in the clear. You might consider putting off your flight until another day.

Traffic

This is just another quick reminder, once again, that you can fly all over the mountains and not see anyone until you get close to an airport. Then you will discover that at certain times of year some mountain airports are very busy. All the traffic, inbound and outbound, is squeezed into the same narrow valley, and you cannot communicate by radio except through line of sight. You can (and should) keep announcing your position and intentions over the radio on the way in, but someone about to depart will not hear you until you have the airport in sight. By then he or she may already be on the roll or have just lifted off.

Finding the airport

Finding an airport in flat country, at least with GPS, is usually an easy enough procedure, although poor visibility can mean you don't see it until you are very close. You fly straight toward the airport until you see it.

Study again your proposed route into Aspen, or whatever mountain airport you chose. Quite probably you will be descending into the valley where you expect the airport to be long before you can actually see the airport itself. You may be planning a descent down a particular side canyon. That means you should have thought of some means of identifying the right valley. You could do this by selecting a GPS waypoint at a certain distance and bearing to the airport. You could also note the position of prominent landmarks on your chart. Doing both would not be overkill.

Once you are headed up the valley, the airport might still not be in sight. Again, studying the map to identify prominent landmarks is necessary to confirm you are flying up the correct valley! Although you should always preserve the ability to turn around, accidents have happened in which selecting the wrong valley and not discovering an airport were the primary cause.

Do not get tricked into descending too low while you are searching for the airport!

Remember that a normal descent for a small airplane is around 1000 ft every 3 mi. I have just been counseling you to begin descending early, but getting too low can be equally problematic. Many, many times I have seen pilots unfamiliar with Glenwood Springs, Colorado, fly all the way up the valley 12 mi south of the field to the town of Carbondale and then descend to near pattern altitude. Next they have to wind their way 12 mi back up the valley over the highway—only to still be completely surprised when they finally find the runway and end up too high to land! This also puts them right in the face of departing traffic and, because they are below terrain, puts them out of radio contact with pilots about to depart. If they were over Carbondale at an altitude of 10,000 ft msl instead of 7000 ft msl, they would have the airport in sight and have a long, straight-in final at a normal rate of descent and normal descent angle.

If pilots flying into Aspen for the first time were to do as the Flight Guide advises and fly up the valley at over 10,000 ft, they would

Finding the Glenwood Springs, Colorado, airport. On a standard glide slope, the author returning home from work at the end of the day is well above Mount Sopris, 20 mi distant. By using a normal rate of descent, the airport is visible from 20 mi away. If you fly lower, following the highway up the valley, the airport does not come into view until you are only 2 mi out, and you arrive right in the face of departing traffic. Shelby Evans photo.

come around the corner 6 mi out on glide slope and in contact with the tower. Many instead fly up the valley, getting ever lower, until at less than 9000 ft msl they are cranking sharp turns up the canyon, can't hear the tower very well, and stumble around the corner only 2 mi out when they finally can see the runway.

If a low ceiling forces you down, then that is how you have to do it. Otherwise you might consider making either mental or pencil tick-marks on the chart noting the target altitude for various points inbound.

Arrivals

Has it really taken that many pages and that much planning just to arrive at the airport? Are we there yet? Not quite. Now you have to get lined up with the runway.

Recalling that because the air is thin at altitude, your actual airspeed is some 20 percent higher than indicated, you will seldom elect to land headed downwind at a high-altitude airport without a very good reason.

If you are approaching from the west but the wind is from the east, you may want to fly a standard traffic pattern: downwind leg, turn base, final approach, and land. If the valley is wide enough, do it. Often, though, nearby terrain is going to dictate how you accomplish this. First let us limit the discussion just to terrain avoidance.

Again, let us take as an example Glenwood Springs, Colorado, which sits at the bottom of a narrow north-south valley and has a north-south runway. The valley is just wide enough to accommodate a standard traffic pattern at standard altitudes and distances from the runway (at least on the west side of the airport). Accordingly, most local pilots will elect to fly a perfectly normal pattern. Entering the pattern from the east, though, they will have to duck down a narrow side canyon, and the field will come in sight when they are less than 1/4 mi out for a midfield crossover. Coming from the north or south, they will just fly up or down the valley. Relatively few aircraft can descend steeply enough (or perhaps more accurately, because some people do manage it, relatively few pilots *want* to descend steeply enough) to enter the downwind leg from the west over the canyon rim. But once the aircraft is in the immediate vicinity of the airport, a standard traffic pattern is possible in any aircraft with a pattern speed of under 130 knots. This all seemed normal to me because Glenwood Springs is where I flew my first solo. I didn't know anything different.

A standard traffic pattern is entirely possible, but pilots unfamiliar with the area or unfamiliar with mountain flying will not fly a standard traffic pattern at this airport. Why not? Because on the downwind leg, 1000 ft above the field and just under a mile to the west, you are 3000 ft below the canyon rim and only 200 ft from the valley wall. When you descend to turn base, you get even lower, and at the point where you actually turn base, you are also below a secondary ridge a couple of hundred yards directly in front of you. On the base leg you are flying directly toward a cliff at least 1500 ft above your altitude and less than 1/4 mi away. Because they are uncomfortable with flying so close to terrain that much higher than the aircraft, most visiting pilots get a little too high on downwind, too high on base, turn final too early to stay away from the cliff, and end up crossing the threshold of the runway so high and going so fast that they probably should go around.

This is very funny to people watching from the ground, but less funny to the pilots because they must make a long, straight-ahead climb out of the canyon before coming back for another try.

Needless to say, even local pilots won't fly a standard traffic pattern into Glenwood if the winds are too high. The presence of so many cliffs and narrow side canyons produces far too much turbulence to fly that close to the rocks. In these conditions, they end up doing what most first-time visitors do anyway. They fly a very, very large traffic pattern 3000 ft above the runway, which puts them 1000 ft above the canyon rim to the east. They enter the valley either north or south of the airport on a 4- or 5-mi final approach.

Avoiding turbulence on final approach

Seldom will final approach in any narrow canyon be made by savvy local pilots on the runway extended centerline. Instead, it will be made along the downwind edge of the canyon in smoother air. Does this sound a lot like the route they choose to depart in the same conditions? Yes it does. It is also the route they would select if they had to go around.

Flying into Telluride on a day with strong south winds presents some problems. Just south of the runway is a high cliff. The runway sits atop the cliff like the deck on an aircraft carrier. Wind deflects up the cliff and forms a rotor right over the pavement. There are deep, narrow canyons to cross on short final approach off either end of the runway. There can be a few days a year, most often in May, when the turbulence can be severe enough to force even airliners to divert.

Aspen has a bluff beside the approach end of the runway. Glenwood Springs and Steamboat Springs have steep dropoffs right at the end of the runway. Most mountain airports have some little peculiarity of terrain near the runway which causes low-level turbulence and wind shear close to the touchdown point.

How, then, can you get in under these conditions? First, admit that sometimes it can't be done (more on this later). Then make your final approach at Telluride north of the runway centerline. A close look at a small-scale map of the Telluride airport will show you that this is an approach lined up over steeply upward-sloping terrain rather than right over the narrow canyon. It will be an approach through less turbulence. Make a steeper than normal approach. This keeps more distance between you and the cliff edge. Fly about 5 to

The author coming home, turning base to final for runway 27, Telluride, Colorado. Notice how close the cliff is to the end of the runway. To stay above the turbulence, I am flying a steeper than normal approach. There are three white and one red PAPI lights left of the end of the runway. I deliberately made the approach right of the runway centerline to stay in smoother lifting air over the hillside. John Kounis photo.

10 knots faster than minimum approach speed, but not faster than that. You can trade off extra altitude for speed if you have to. Fly the approach with very precise speed control so that you can select your round-out point precisely. Pick a point as far as possible from any cliff edge or gully. Use no more than 20° of flap, or no more flap than you can go around with.

There are pilots who advocate a much faster, flatter than normal approach in these conditions. The idea is that you blast through all the turbulence on final approach and then bleed off all the extra speed in a long round-out and flare over the pavement. I do this in heavier planes, but I don't like this idea for a small, light aircraft for two reasons. I would rather make a high, steep approach because there is more room to go around and a small plane accelerates quickly. You can trade altitude for more airspeed if you need to. The steep approach keeps you farther away from the terrain that caused the turbulence. Second, the steep approach still allows for a normal touchdown speed. The more slowly you are going on the ground, the easier the airplane is to control. I suppose that part of my preference comes from the fact that I learned to fly and frequently fly into very short runways, where the fast approach won't work. But as with many other theories, neither is always right. Experience will tell you which method you prefer for your aircraft.

All mountain airports located in the bottom of narrow valleys can experience very strong turbulence accompanying strong winds. *All* mountain airports in any kind of valley can experience severe turbulence during the descent into the valley. This should suggest to you that although you might be able to fly over the mountains on very windy days, you might not be able to land anywhere. *Despite that, the guiding principle for selecting an approach path to avoid turbulence is simply to avoid the lee side of steep ridges and valley walls.*

Going around

Keep it in mind that you don't have to land just because you are on approach. You could go around. You could even go somewhere else.

Jessica McMillan provides an example for all of us to profit by. Jessica McMillan was formerly the airport manager in Glenwood Springs, Colorado. For some reason, although there were two excellent mechanics in Glenwood, scheduling difficulties sent Jessica's aircraft to Montrose, Colorado, for servicing. Some aircraft were ferried down, and two aircraft headed back, carrying the ferry pilots.

The first aircraft in was a Cessna 182, flown by a very macho, newly minted instructor. Because of turbulence he was too fast and too high on approach to land on runway 32 and had to go around. As reported to me by one of his guilty passengers, his two private pilot/passengers immediately teased him along the lines of "Gee, you're supposed to be a commercial pilot, and you can't even get the airplane to the end of the runway." Clearly flustered, he did a 180° turn back to runway 14. Unfortunately, he turned around first and reduced power second, so once again he was too fast and too high and had to go around.

At this point Jessica arrived in a Cessna 172 with one passenger, made an approach, and decided early to go around because of the turbulence.

The macho instructor once again endured the taunts of his passengers, headed back for runway 32, and was once again too fast and too high. Unfortunately, although this approach was better, it was far from perfect. In his mind, however, he was firmly committed to land rather than be embarrassed for a third time in front of his pilot/passengers. When they realized what was happening, both passengers began shouting at the same time for him to go around.

There was no way a commercial pilot was going to take their amateur advice! At midfield the aircraft was not yet on the ground despite a moderately strong headwind. The macho pilot forced the aircraft down into a wheelbarrow landing on the nosewheel, locked up the brakes, wore flat spots on both brand-new main tires down through the cord, and slid off the end of the runway into the gravel, coming to a stop within 10 ft of the barbed wire fence that separated the airport from a nearby public road. Both passengers were very quiet until they got out of the plane.

The aircraft had to be flown back to Montrose another day for two more tires.

Just what was going on here? On the surface, it just seems to be a case of a pilot not very good at landings. That, to a mild degree, is still true. He has run off the runway at least three more times that I know of, and no doubt he keeps other incidents as quiet as he can. The short field landing remains his personal Achilles heel, but going around for another approach would address that. This pilot displays a personality trait that in the wrong circumstances could be very dangerous to his passengers. But another equally important factor is the role that my buddies riding with him played. Their good-natured teasing was not particularly compatible with his fragile ego. In effect, they were putting pressure on him to do something. They didn't really mean to pressure him, but *he* felt the pressure quite strongly. His desire not to be embarrassed in front of experienced *private* pilots when he was a much less experienced *commercial* pilot put tremendous psychological pressure on him. Without that pressure, perhaps he might have elected to handle the situation differently.

In the meantime, Jessica made another approach, didn't like the way it was working out, and went around again. By some accounts, Jessica made as many as eight approaches and went around each time. (By her own account, the total was six, but eight makes a better story.) This is not so bull-headed as it might sound. In varying conditions, getting in might depend on the good luck of being on short final approach just as a lull came through. Eventually, though, Jessica elected to fly to the nearby Rifle Airport in a broad valley, with the runway aligned with the strong west wind. An hour later when conditions improved, she flew back in without incident.

Even if not being embarrassed is greater motivation to you than not risking your passengers' lives, almost no one remembers that Jessica made six (or eight?) approaches and did not land. Everyone remembers that the professional pilot ran off the runway and trashed two new tires.

When instructors do dumb things, there is a snowball effect because students tend to pick up on what the instructor *does* as much as on what he or she says. We all hope that given a choice between these two examples, the students present subconsciously decided a humble go-around is better than a heroic botched landing.

Another example but one with a tragic outcome.   The crash of a Gulfstream jet on a night winter instrument approach into Aspen, Colorado, has already been alluded to. When first released, transcripts of the cockpit voice recorder had been edited, or perhaps censored, by the FAA. In these transcripts it is very apparent that the pilots had concerns about making a night instrument approach into that airport, and they understood perfectly well what the special dangers were. They are discussing Rifle, Colorado, as an alternate, and say if X happens or Y doesn't happen, then we're out of here and over to Rifle. A plane before them misses the approach. It is not hard to tell that neither pilot wants to do this particular approach. But they do it anyway, lose sight of the runway in the snow, and crash into a hillside. All 17 people aboard die.

A few people have since told me in confidence that they have heard the unedited tape. The editing changes the story considerably. On the unedited tape, there is a sort of passengers' representative or company spokesman of some kind who continuously comes up to the cockpit to check on the progress of the flight. He delivers a constant stream of "We really have to get in to Aspen," "We really can't divert for the van ride," and "I've been into this airport plenty of times. If you guys can't do it, maybe we should be hiring somebody who can." Now you would hope that the pilots could just tune that out, but tuning it out completely is asking more than normal human nature will provide. No doubt the pilots believed that if the situation got too dangerous they would overrule this guy, but the reality is that, faced with the possible loss of their jobs, the pilots tended to reassess the dangers in a way that would please the guy who could have them fired.

Among others, the airlines have become much more aware of this kind of pressure, and most are adopting a new policy. If either pilot declares he is unhappy doing something, then it isn't done. The more conservative judgment call overrules any other. In this instance, the junior first officer can overrule the senior captain, and there is to be no debate about it.

This might be a reasonable rule for private pilots as well, however hard it might be psychologically to follow.

Landings

Landings at high altitude airports require no special techniques not also required for low-altitude landings! I am including a discussion of landings for two reasons. First, you may have read or been given advice which directly contradicts the previous sentence. Second, because they are sometimes small and located in marginal surroundings, certain low-use mountain airports are very unlike the airports that typical aviation textbooks discuss.

Landing speed

You may have read or been told that because the air is thinner at high altitude, airplanes have to be landed at higher speeds in order to maintain enough lift. This is entirely correct. The air is thinner. In order for enough of it to flow over the wing and produce adequate lift, the plane must be going faster.

What I disagree with vehemently is the next bit of advice which follows that advice. You may have read or been told to add 10 or 20 percent to your approach speed. The basis of this advice is that true airspeed increases by about 10 percent at 6000 ft msl and 20 percent at 10,000 ft msl. You need to be flying a true speed about that much higher at those altitudes to achieve similar performance. Therefore, the advice goes, add 10 or 15 knots to your approach speed.

Don't do it! Since the air coming into your pitot tube is also thinner at altitude, your indicated airspeed is in error. It is in error by precisely

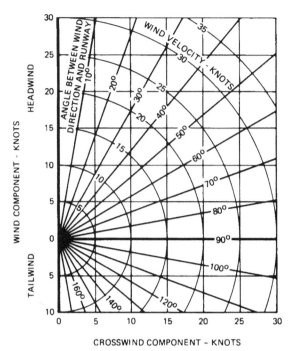

CROSSWIND COMPONENT – KNOTS

This crosswind component chart provides at a glance the effect current wind velocity and direction will have upon your takeoff. For example, a 25-knot wind at a 30 degree angle to either side of the runway heading has the same effect as a 13-knot crosswind. For more details on the effect of wind on your airplane at takeoff or landing, consult aircraft owner's manual.

the amount needed to correct your true airspeed for higher altitude. *Your airspeed indicator automatically corrects your speed for you. Fly the same indicated airspeeds called for in any given phase of flight at all altitudes!*

What will seem different is this: Your speed over the ground in a high-altitude landing will be higher than it would be for the same conditions at sea level. At 9080 ft msl at Telluride for most small aircraft, a calm wind landing will look and feel a lot like a sea-level landing with a 10-knot tailwind. Any prudent pilot flying a small aircraft would normally elect not to land with that much tailwind.

At high altitude, all your landings are like that! While you might not do anything different, obviously if things are happening 10 to 20 percent faster, it is clearly advantageous for you to be able to do them 10 to 20 percent better. If you learned to fly in those conditions, you never realized you had learned anything special. If you learned to fly at sea level, you are going to make an adaptation equivalent to stepping up to a slightly higher-performance aircraft.

Obviously the slower the aircraft is going when it touches down, the greater control you will have and the less wear there will be on the tires and landing gear. One way to achieve a slow touchdown speed is to always land with full flaps. If the wind were always calm, you might do that. I don't teach my students to always land with full flaps, in part because with the low engine power available at high altitude, it can be very difficult to slow or stop a descent made with full flaps if you encounter sink. It can be extremely difficult to go around. If your particular aircraft is one which should be landed in a crosswind with no more than 20° of flap, then at altitude *you still follow the same procedure.* In strong winds or turbulence, land with no more flaps than you can go around with, or at least don't go to more flaps until you are certain the runway is made.

Certain aircraft, and certain pilots, always land with zero flaps in a crosswind at sea level. If that is the case for you, keep doing it that way in the mountains. My personal preference, depending on the make and model of aircraft, is to use some flaps even in a strong crosswind to get the touchdown speed as low as possible. Then I instantly get rid of the flaps as soon as I touch down, in order to get rid of excess lift.

Speed control on final approach

The operator's manual will have recommended speeds to fly on short final approach. At sea level it is possible to be fairly lax in adhering to these speeds. In the dense air of lower altitudes, any light aircraft will decelerate rapidly once the power is pulled off. A slightly too fast speed over the threshold of the runway will have minimal effect on landing distance. A too slow speed can be corrected with a burst of power. At higher altitudes, you have greater inertia because the airplane is going roughly 20 percent faster at the same indicated airspeed, while drag is less in the thinner air. Slightly too fast will result in a much longer landing. Conversely, if you get a little slow at low altitudes, lowering the nose quickly restores airspeed. At higher altitudes, there is a longer delay in regaining speed because although drag is less, your aircraft has to accelerate to a higher *true* speed to obtain the required *indicated* speed. Also, the aircraft has all the momentum it always had, despite the thinner air.

Excellent landings at any altitude require a high degree of precision, but *at high altitudes, very precise control of airspeed is more critical than at low altitude.*

Uphill/downhill runways

Because there is not always much flatland available, and because they serve communities too small to afford bulldozing out a big, flat place, many smaller mountain airports have a pronounced uphill slope one way and downhill slope the other way. This slope introduces some special considerations.

Unless you are forced to do otherwise, always elect to land uphill and take off downhill.

Landing uphill can greatly reduce your ground roll. Probably 10 percent per degree of upslope is a conservative figure. The actual amount of reduction may be greater with lighter, slower planes and less with bigger, heavier ones.

Landing on a steep upslope has a visual trap that can cause problems. You are used to a certain visual picture of a normal runway on a nor-

The very steep upslope allows ordinary single-engine aircraft to use this runway even though it is only 600 ft long. The runway is clearly one way in, one way out, and far too steep to go around. John Kounis photo.

mal approach. The same visual picture when you are on approach to a steep uphill landing means you are flying a much flatter—perhaps too flat—approach. You risk the low and slow trap described below in "Short Field Landings," and correcting your glide path could actually require a fairly steep climb. Make very sure that you are coming in to land at a normal descent angle and descent rate! Better yet, if you are a low-time pilot contemplating this kind of landing, get an instructor familiar with some steep uphill field to show you how to do it.

Most pilots advocate adding about 5 to 10 knots to your final approach speed for a steep uphill landing. The reason for this is that the round-out is a tighter pull-up than normal, and slightly greater speed produces greater lift to facilitate this. While a touch of extra speed may make the procedure easier, it is not absolutely necessary. Neither does it do much harm—the plane will slow down very quickly once it is rounded out because it will be flying uphill rather than level.

A botched uphill landing—too many good ideas. As with any procedure, it is possible to go a little too far. A pilot based in Montrose, Colorado, landing on a somewhat short, steep grass runway near Telluride in a Cessna 182 elected to fly a fairly steep approach

This view shows the 89-ft-deep dip in the center of Telluride's runway—you can land going downhill in either direction! This pilot elected to deliberately land long on the upslope (probably at the request of the photographer). John Kounis photo.

(good). Also he elected to carry a little extra speed (also good). Finally, he decided to aim at a touchdown point slightly farther up the hill than normal because he believed the lower end of the runway was bumpy (also reasonable, but not really as safe as using all the runway available). The week before, this same pilot had come in too slow at another airport while flying a Cessna 150, stalled it in, and broke the nose gear. This time he would not repeat that mistake—he would deliberately be faster, higher, and longer. Doing all these seemingly good things at once was simply too many good things.

He hit the ground very hard on the nosewheel at a point past midfield, locked up the wheels, and slid to a stop going sideways in the gravel at the very top of the runway. In this case, the very top of the runway is also the very top of the hill. Keep going and you fall off a near cliff 300 ft into a gully.

The moral here is that when you deviate from your normal procedures, just change one thing at a time, and only change it a little.

Downhill landings

A very strong head wind might cause you to land downhill. Landing downhill can increase your ground roll by at least 15 percent per degree of slope, possibly by much, much more. An even bigger problem than increased ground roll is increased floating distance before you touch down. Remember that a standard glide slope is only a 3° slope. On a normal approach to a 3° downhill runway, you might never touch down at all! A good, strong upslope wind could almost convert this float to soaring and keep you in the air indefinitely. If you manage to fly the approach at precisely the correct approach speed and make a relatively steep approach, then the increased floating distance will be minimal. If you are as little as 10 knots too fast, then your float could exceed the entire length of some very small, steep runways. Landing on a steep downslope has little to recommend it other than that it is easy to go around! Before you attempt a downhill landing on a steep runway, seek the advice of people who have landed on that particular runway in the same type of aircraft. Alternatively, just wait until the wind allows you to land going uphill.

Short field landings

Before you land on a short field, high or low altitude, determine whether the field is long enough to use by going to the takeoff and landing tables in the operator's manual. For nearly all nonturbocharged aircraft, landing distance at high altitude is far shorter than takeoff distance. For most small aircraft, landing distances at 10,000 ft msl are about 20 percent longer than those at sea level, while takeoff distances triple. Only a little planning may save you from landing somewhere you cannot fly out of.

Older textbooks and older pilots advocate landing on a very short field by coming in very low, very slowly, with full flaps and power on to hold you in the air; then chopping the power just as you cross the runway threshold to drop right in. A low, flat approach with considerable power can be made as much as 10 knots or more below your slowest power-off approach speed. This is a very effective way to make a very short spot landing at sea level, but not necessarily a safe method. You are right on the edge of a stall, with almost no reserve ability left to either climb or recover. At high altitude, it is very unsafe indeed, and in the mountains it is particularly risky.

At high altitude with less engine power available than at sea level, it is very easy to get the aircraft behind the power curve. In this situa-

Short field landing. The pilot has configured his aircraft for the shortest possible landing by using full flaps and approaching at the lowest safe speed possible. He is wisely using a very steep descent angle with power slightly on, so that if need be, he could trade the extra altitude for more airspeed. John Kounis photo.

tion, you are flying very slowly with the nose held very high. Just a tiny percent slower, and adding power no longer produces a climb. If you continue to hold the nose up, you can gradually feed in power until you are at full power but losing altitude. Your only escape is to lower the nose; but if you are already very low, doing that would cause you to hit the ground. Reducing the amount of flaps would allow you to climb; but again, if you are already very low, reducing flaps would cause a momentary settling which would make you hit the ground. Low and slow with flaps is a dangerous trap, particularly at high altitude and particularly in the mild turbulence that is often caused by rough mountainous terrain.

If the runway is so incredibly short that this technique is required, then there is not enough distance to take off again anyway.

For high-altitude, mountain, short field landings, the method found in the FAA *Private Pilot Practical Test Standards* works just fine. Fly the last part of the final approach with full flaps at the very minimum speed allowed by the operator's manual. Make a somewhat steeper than normal approach so that you can convert altitude to airspeed if necessary. Pick a round-out point just before the end of the field. Be prepared to add full power and go around if it does not look as if you are going to touch down at the very end of the runway.

With practice, most small aircraft can be brought to a stop in a very impressively short distance by using this method. Nothing more radical is required to land in a space much shorter than you need to take off!

If it becomes your personal goal to land on shorter fields than everyone else, for example, if you need to keep your plane at home and fly in and out of your driveway, then a power-on very slow approach can shorten the distance. At least at high altitudes and in the mountains, a slow, power-on *steep* approach *might* work. A power-on *slow, flat* approach is a recipe for disaster. Certain types of aircraft such as Piper Super Cubs and Aviat Huskies can use this technique very effectively for off-field landings. A Helio Courier, which has leading-edge wing slats that deploy automatically at very low speeds, remains controllable at speeds substantially below its stall speed. It in fact does not stall at all in the conventional accepted sense of the word. A 4000-lb fully loaded six-place Helio Courier can be flown down power on, essentially stalled at speeds as low as 24 knots. Given any wind, it almost appears to hover in to a no-ground-roll landing. Special training from someone who uses that technique frequently in that type of aircraft is a must.

Rough field/soft field landings

If you fly regularly in the mountains, you can expect to become very proficient at soft field landings for two reasons. First, since the mountains are not very populated, many smaller places cannot afford to pave their runways. Many places are so small in fact that they have not even managed to get their runways on the sectional charts. Second, even the largest mountain airports see plenty of winter snow. Even if the line crew has been plowing snow all night long, you cannot expect to be guaranteed a dry runway in the winter.

Soft field landing techniques are discussed shortly.

Check the field conditions before landing. Somewhere you must have learned not to land without first determining whether the field is suitable. For a typical paved airport, this involves no more than checking the wind direction and runway length. For any other type of airport, the process can be much more involved. If you plan on landing at a dirt runway which is in regular use, it is prudent to at

For this short, steep uphill landing, the pilot is correctly making a normal, not a flat descent. John Kounis photo.

The fact that this descent is not flat is a particularly good thing, because otherwise he could not clear these trees and still land at the very end of such a short strip. John Kounis photo.

least fly over the field to check conditions. If there is any question, it is best to make a low pass of the runway to take a very close look. This could tell you about such occasional hazards as ruts and mud holes or rain gullies which can sometimes be found on any dirt field.

Fry Canyon Motel, Utah. Tall bunch grass makes this runway look rougher than it is. Ordinary nosewheel aircraft do just fine on this kind of runway provided good soft field takeoff and landing technique keeps the nosewheel from touching the ground, except when the plane is moving as slowly as possible. Fry Canyon, near Natural Bridges National Monument and Lake Powell, is officially the most remote motel in the state of Utah and an occasional charter flight destination.
Fletcher Anderson photo.

Prairie dog holes bigger than most nosewheels plague some western Colorado fields. Few unpaved fields are suitable for use following a heavy rain or a wet snowfall or recent snowmelt. Snow depth and snow conditions are very easily misjudged from the air, and special techniques are required.

If the field is overgrown with grass, known to be soft, or just very infrequently used, it can be necessary to determine field conditions from the ground before landing. A good procedure is to speak with someone who has used the field recently. Obviously walking the length of the field is best, but this is not always practical. Walking the field could show you that landings might be practical along only one side, or at only one end of the field.

If neither of these options presents itself, landing somewhere else first is not a bad idea at all. *Making a very low pass of a field like this will not be sufficient to determine landing conditions for a typical nosewheel small aircraft,* although it might be fine for a tail wheel aircraft intended for off-field landings. Standard small aircraft nose gears are not strong enough for repeated, very rough field or off-field landings! Aircraft routinely used for this purpose have tail wheels or beefed up nose gear.

Soft/rough field landing technique.   The problem encountered when landing on a soft or rough surface is that the main wheels can be expected to abruptly encounter considerable rolling resistance or even dig in. Weight will then be suddenly transferred forward. With a nosewheel airplane, this abrupt slamming down of the nose could cause a loss of control or even damage the gear. Tail wheel aircraft have the main gear mounted farther forward to take this kind of shock. In an extreme case, though, a tail wheel airplane can nose over right onto the propeller.

Avoiding either of these mishaps involves the same basic technique. The idea is first to touch down while going as slowly as possible and as softly as possible. This is accomplished by using full flaps and holding the nose as high as possible, floating the airplane in the air as long as you can until it just can't stay airborne any more. This is the way you *should* be landing all the time anyway, except that you don't always use full flaps.

Adding the tiniest possible amount of power once you are rounded out will perform an absolute miracle in terms of how softly you touch down. It will also allow you to hold off touching down slightly longer, as you will be able to attain a higher than normal nose-high attitude and with it a lower speed. If you touch down with the power still on, it will be much easier to keep holding the nosewheel off the ground after the main wheels touch.

Once the main wheels have landed, *leave the power slightly on,* hold the yoke full back, and let the nosewheel settle on by itself. Power on or power off, this is the way you land anyway, isn't it? You don't suddenly relax all the back pressure on the yoke and let the nose slam down, do you?

As you roll out and as you taxi, always continue to hold full back pressure on the yoke to keep pressure off the vulnerable nose gear.

Landing in snow.   The actual *landing* in soft snow is simply a standard soft field landing. The hazard of a soft snow landing is getting stuck.

Turning once again to the teaching technique of using someone else's stupidity as a bad example to profit from, I will now recount an example of a snow landing nearly gone very bad. While it is generally my practice to avoid naming the guilty in the hope that they might have now reformed, I will make an exception in this case. This is me personally doing something dumb in an airplane. I hope I have reformed.

In December 1996, I had a succession of knee surgeries and had to give up skiing. I was also working two full-time jobs. On my few days off it was my habit to load my bicycle into the back of my Cessna 172 and fly out of the snow country into the Utah desert to bike ride. One of my favorite stops was Sand Wash, which is not on the chart, but is about 55 air miles upriver from Green River, Utah. Sand Wash is used in the summer to drop off river trip passengers for Desolation Canyon, but sees no humans whatsoever in winter. *Desolation? No humans? Winter? Off the chart?* If not the middle of nowhere, it is at least very remote.

On this particular trip, I was surprised to find the ground covered with snow. Instead of landing, I flew two passes over the runway about 10 ft off the ground to check things out. Animal tracks and grass showing through the snow were enough to tell me the snow was only 2 in or so deep—no problem for landing, and certainly not too deep to bike ride either.

I set up for and made a regular soft field landing. *Surprise!* Powder snow billowed up right over the top of the windshield! Adding full power kept the airplane moving at something just slower than a walk.

I did not dare come to a full stop. My only hope was to keep moving the entire length of the runway to the very end, keep clawing forward around a turn until I was going back the other way up my tracks, claw around another turn at the other end of the runway, and repeat the process. After a couple of passes, it no longer required full power to keep taxiing, but full power produced nothing like the speed needed to take off. In fact it took no less than six passes of the runway to pack out three ruts over a foot deep from which I might get airborne again. Naturally enough, my track was not precisely straight, and on my first two takeoff attempts I bounced out of the packed track in a slight dogleg at midfield. Only on the third attempt did I manage to get airborne. Bike riding was out for that day anyway.

Had the snow been only a couple of inches deeper, or had it been something else other than very dry, light powder, I could have been truly stuck. I would have been forced to foot-pack a track for each wheel the full length of the runway before I could use the plane itself as a very inefficient snow cat for more grooming. With a leg just out of surgery, this could have taken some time. Very wet or wind-packed snow might have really and truly stranded me in a location where there is no cell phone coverage and no radio contact with anyone. I could have camped until the first river trip arrived in June.

Alaskan pilots face this problem all the time, but they treat it differently than I did. They deliberately make the first landing a touch and go, adding full power just as the wheels brush the snow. They then make a succession of touch-and-go landings, each time "dragging" the strip a little longer and a little more firmly so that they manage to pack out a runway without slowing below takeoff speed until it has been fully packed and tested. Only then do they actually do a complete landing. The idea is not only to ascertain if landing would be safe, but also to pack the runway for the takeoff *before* committing to being on the ground.

Professional Alaskan bush pilots will readily admit that even they cannot accurately judge snow conditions from the air. They always drag the field a few times even when landing on skis. Even then, even for them, even with bush planes, this is still a risky procedure, and they sometimes get stuck or occasionally end up with the plane on its nose. They frequently have to resort to rocking the plane to get it moving, or have to have a passenger get out and give them a push to get started. Then they taxi by again and the passenger has to jump into the moving plane. Despite all these precautions, tales of their getting stuck are common every winter.

Landing on ice.   *Landing* on ice requires no special techniques; neither does it pose any unusual difficulties. After your first winter in the mountains, you will have done it many times. Difficulties arise after you have landed and you are going slowly.

Extremely cold ice provides reasonable traction. Wet ice, such as you encounter as the sun begins to warm ice frozen the night before, or

Wet spring snow. This late spring snow will melt off soon, but right now there is very poor braking action on the snow patches but good braking on the wet pavement. The idea here is to maintain control by touching down as slowly as possible and then using minimal braking.
Shelby Evans photo.

such as you encounter after wet snow is heavily packed out by other traffic, can occasionally have traction approaching zero. Lucky you if you are landing at an airport where unicom or the tower can give you an advisory on traction reported by other pilots. Otherwise you are on your own.

When you first touch down, you can expect to have no trouble maintaining directional control because there will be plenty of airflow over the rudder. As you slow down, rudder effectiveness is lost, but tire traction may not take over. It is important, if you suspect this may occur, to make your landing very straight and very precisely down the center of the runway.

Hold the airplane off as long as possible before touching down, as it will slow better aerodynamically in a nose-high attitude than it will with braking on the ground. While I am sure you routinely grease the airplane so smoothly that you can scarcely tell when the wheels touch (don't we all?), that can allow the wheels to just start sliding instead of rolling. Bumping it slightly on ice is better because the initial contact will get the wheels rolling. After that, as with your car, apply the brakes gently. If the wheels skid, ease off the brakes until you stop skidding, then reapply the brakes. Sometimes it seems you are just pumping the brakes on and off for hundreds of yards.

You can stop effectively, it just takes more room and you will have to taxi very slowly afterward.

In the very unfortunate event that you have landed but are clearly going to slide off the end of the runway, your best bet is probably to skid off into the snow straight ahead. If the ice is very uniform and very slippery (for example, if you landed on a frozen lake rather than at an airport), you could consider deliberately ground-looping and then adding power while sliding backward to stop. The obvious risk is that you will find enough traction when going sideways to trip and wreck the plane. Quite likely the best thing to do is always to just run off the end of the runway. The ground there is usually smooth, and the deep unplowed snow will stop you.

Some examples.   Some of these examples involve larger aircraft, but the principle is the same—just the speeds are different.

The pilot of a twin turboprop coming back empty from Montrose to Telluride was feeling bold. The temperature was barely above zero, and the snow made that groaning noise always associated with severe cold weather. Snow this cold actually provides pretty good traction, so the pilot was inspired to try to make the first turnoff landing to the east. Hard braking locked up the wheels, and the aircraft gradually slid the left gear off the left side of the runway before the pilot, still focused on making that first turnoff, attempted to steer by putting more reverse thrust on one engine than the other. He overcorrected and careened off the right side of the runway. Once off the runway in deep powder, the aircraft did in fact come to a stop just barely beyond the turnoff. Presumably if the pilot had just tried to stop normally and under control, he would have had little difficulty stopping in the remaining 2500 ft of runway or the overrun space beyond. If he had released the brakes after the skid began, or if he had used equal amounts of power on the two engines, or if he had just eased off reverse thrust on both engines for a few seconds and let the rudder straighten things out, probably he could have straightened the plane out. In his defense, this particular plane is not easy to control on the ground. But that should suggest to him that he should try to land it more conservatively.

As luck would have it, the right-hand propeller scrambled a steel runway light pole. All three propeller blades had huge chunks missing, requiring replacement and a propeller overhaul, which on that type of aircraft runs over $25,000. The sudden engine stoppage required a gearbox overhaul at a cost of over $35,000, and the mandatory engine overhaul cost exceeded $100,000. Also this came on the first day of Christmas vacation when the aircraft should have been busy flying daily charters.

The bravado exhibited by almost making the first turnoff impressed no one but the mechanics who came to fix the plane. They treated him to lunch several days running and no doubt hope he keeps flying other people's expensive planes!

Coincidentally, this same aircraft had spent 4 h the previous day on the ramp in Montrose, Colorado, having the blades of the same propeller filed smooth. The same pilot had run off the edge of that runway that day as a result of taxiing too fast and picked up some debris with the propeller.

Bad judgment, not lack of skill, is at play here.

Scarcely a week later, a much larger and heavier Canadair Challenger coming in from Missouri landed long and fast on the same runway at Telluride just as the temperature rose toward the melting point, making for a very slick surface. The very large, heavy aircraft slid straight ahead off the end of the runway. The pilots then turned slightly left where the terrain goes uphill to help stop, coming to rest a good 100 yd beyond the end of the pavement in deep powder snow. Although it was not extracted until 9:30 p.m., the aircraft was undamaged.

During the years I have been working at Telluride there have been more snowy winter days than anyone outside the ski company's marketing department cares to count. Twice that I can recall, after touching down with over a mile of runway in front of them, heavy aircraft just kept on sliding. In those instances, the pilots simply did their best to keep the aircraft under control and pointed straight ahead. They ran off the end of the runway and into the safety overrun area and came to a stop in deep snow. The aircraft had to be towed out but were undamaged. The pilots were deeply embarrassed, but no more than that. During the same period, various aircraft attempting to stop short have had various amounts of minor damage.

At most airports, even at many remote backcountry strips, there is a smooth area off the end of the runway intended as a safety overrun. The ground beside the runway may be cut by berms and ditches.

A small aircraft might have to choose between these two alternatives on a short field and would be risking a propeller strike in either case, but straight ahead reduces the risk of tipping over. On a long runway in a small plane there should be no trouble at all on the slickest ice, provided the pilot does not try to do anything fancy. Ice or not, the aircraft will eventually just roll to a stop even without any braking action.

At just the right (or wrong) instant late on a spring afternoon, I was in a Cessna P210 following Todd Wilson in a Cessna T206 back to Telluride. The very same wet runway as in the other examples had just frozen, and the ramp was so slippery that several people fell down just trying to walk across to their planes. Telluride is unique in that you can land going downhill in both directions—the center of the runway is 78 ft lower than the ends. You can land downhill on ice. How quickly can you stop from that one? Todd landed to the west, touched down right at the end of the runway at the slowest possible speed and just kept going. At the first taxiway he radioed "braking action reported as zero by a 206." At the second taxiway he reported "braking reported imaginary by a 206." He just kept on sliding, but the runway is 6800 ft long. He kept pointed straight ahead. Eventually the aircraft simply glided to a stop at about midfield, and he taxied in.

Since there was nothing else equally amusing going on that afternoon, the entire line crew assembled to watch my landing. Sure enough, although the landing itself was about as good as I can normally make it, slowing down afterward just wasn't happening at all. But again, this is no problem in a small aircraft. The Cessna P210 I was flying has a gross weight of 4000 lb, that is, about the combined weights of two 172s flown with two people aboard each, but still at the small end of the aircraft scale compared to the big jets that routinely land here. It slid and rolled and slid and rolled and glided and rolled....But I just worked at staying on the runway centerline and let it keep going straight, just as I had to do in my car later on the drive home from the airport. Eventually on a 6800-ft-long runway, a small airplane just runs out of inertia and comes to a stop. It required a very boring afternoon indeed for the line crew to be entertained by that.

Trying to stop short would have been impossible that day and could have introduced all sorts of control problems. Landing on a 2000-ft-long runway would not have worked. However, just landing on wet ice on a long enough runway is so uneventful that in hundreds of these landings every year, nothing remarkable ever occurs unless the pilot induces it.

Another tale of ice and snow from the great white North.   For many years, John Spencer was a member of the U.S. national cross-country ski team before returning to his native Alaska and taking up a career in flying. Coincidentally, I had made some of his skis when he was racing, but first learned he was a pilot when he dropped off kayaker Barney Griffith in his kayak for the first descent of the Susitna River. John flew up the canyon 5 ft above the rapids, landed on floats in midstream. Barney unleashed his kayak from the floats, climbed in, and paddled off. John turned the plane around and headed downstream for the rapids, using the bump from the first big wave to get airborne.

John later flew for a lodge, dropping skiers high up in the snow-fields. Early in the season, there was snow at the lodge, and his plane was on skis. Later he changed over to those huge balloon tires that are almost an Alaskan trademark. Finally, though, he put on floats to drop off fly fishermen, but still continued landing skiers on the snow, using the floats as giant skis. The slope where he was dropping the skiers was quite steep. John would land uphill and turn across the slope to stop. The skier would get out and then lift the tail of the plane up the hill to give John a start rolling back down. Eventually John realized that the slope was steep enough that he could coast downhill fast enough to actually take off without starting the engine! On skis, on balloon tires, *and on floats,* he would just slide downhill faster and faster until he became airborne. He would then glide down and land without ever starting the engine. With floats, he was able to land in the lake just the right distance from shore to coast slowly up to the dock.

John and many other Alaskan pilots also land on the water with those big tires they all love. At its wheel landing speed of about 40 knots, a Piper Super Cub, and many other light airplanes, will aquaplane along the surface of the water on the big tires. You chop the power just as you hit the beach. You pray that you are going fast enough when you hit the water on your takeoff roll to keep planing until you are going fast enough to fly.

Off-field landings

Off-field landings are beyond the scope of this book, but some words of caution are in order with regard to the mountains. Alaskan pilots routinely land on snowfields, in meadows or on the open tundra, on gravel bars, in the shallow water of the river itself, or even on deep water and water ski on their fat tires until they can stop on the beach. They can land seemingly on a postage stamp anywhere the ground even vaguely approaches being flat. They do this naturally enough in planes built or modified for that purpose. These planes can fly very slowly and have very robust landing gear. More importantly, except when landing on snow in the mountains, they normally do this at lower altitudes.

Trying the same thing at high altitude in the mountains means that your speed over the ground is perhaps automatically 20 percent

The biggest difference between a dirt road and a dirt runway is that the road is a lot longer. It is a very good idea, though, to check carefully first to see if there are any utility lines or speed limit signs you might run into. Shelby Evans photo.

higher, so the landing is effectively twice as rough or more. Stopping distances are longer. Takeoff distances are over twice as long, if not over 3 times as long, and takeoff occurs at a 20 percent higher ground speed. The chances that you will damage the plane are several times higher. Those Alaskan pilots who routinely land off field at high altitude in the mountains generally do so under very limited specific and controlled circumstances. These are typically uphill landings, preferably on snowfields or glaciers, or on smoother surfaces that are close to being grass runways. They do so at times when they can be relatively sure the process will not be complicated by unacceptably high winds or turbulence. They check out the landing area carefully before they try to land—and they still sometimes damage their planes and get stuck.

Off field, mountains, and high altitude can be one factor too many, and nearly all pilots leave the idea alone. Helicopters were invented for that purpose.

Should you elect to ignore my warnings, the technique is the same as any soft field landing. The goal is to touch down while going as slowly as possible. The advice to check the field very carefully on the ground before attempting a landing cannot be overemphasized.

7

Only in the mountains

Unconventional operations

Thus far, this book has been a discussion of the things you need to know about flying in the mountains that are not covered in standard texts. The subjects in this final chapter could be considered as appendices. They cover some of the things you probably won't need to do.

Flight close to terrain
Search and rescue, aerial photography, game spotting

All these are variations on a theme. That theme is that of necessity you find yourself flying much closer to terrain in places you would not normally fly in atmospheric conditions you would not choose. Why is this dangerous? Well, you could hit the terrain!

There is very little formal training available for this sort of flying. One of the Civil Air Patrol's primary missions is to provide flight instruction for its members, and this includes some low-level training. There are a very few schools offering agricultural flight training, i.e., crop dusting. The military teaches ground attack to a few pilots. A few sheriff's department search and rescue groups do a little training. For all practical purposes though, you are on your own in learning to do this. Proceed very, very gradually, and take advantage of any help another pilot experienced in this may offer.

I received invaluable training in this kind of flying from a pilot who is an absolute master of his craft. His explanation is that many years ago he had a couple of gold mines down in old Mexico and flew back and forth between them and Colorado, never needing to file a flight plan and landing on the roads near the mines. He also landed at night on roads in Colorado on occasion. Now it takes only a little bit of deductive reasoning to figure that it would be easier to haul gold in a truck than in a Cessna 206. It would also be easier to just file a flight plan, clear customs, and cruise at least 500 ft above the ground. It takes quite a bit of skill to routinely fly below treetop height at night over rugged terrain. In retrospect, these particular skills would be most useful for hauling something other than gold out of Mexico below the radar.

I received additional training from a pilot who flew similar missions for Air America in southeast Asia.

What they taught me has certainly kept me alive and in the process helped me be a factor in saving the lives of some other people. Thanks, even though I know you don't want your names mentioned.

It should be self-evident that these forms of flying are inherently much more dangerous than normal operations. That danger is reflected in sometimes very substantially higher accident rates. Yet the extra degree of danger is still greatly underappreciated. All young hotshot pilots assume they will be good at this. Doing it right is much more demanding than it seems. Every year hunters scouting for game hit the ground. In Alaska this is called the *moose stall*—the pilot gets low and is slowly maneuvering to see a moose and stalls and spins climbing out. The Civil Air Patrol has lost trained pilots involved in search missions. Hollywood has lost photographers. Crop dusters who do this day in, day out 8 h/day still frequently hit the ground. Because of the kind of terrain you fly through in the mountains, it should come as no surprise to learn that these tend to be very serious or fatal accidents. There is a very admirable human reaction when someone else's life might be in danger to take excep-tional risks in the hope of saving that person. Yet the inherent risks in the type of flying needed for a search do not go away because of the humanitarian motives of the searchers! The first rule of search and rescue is not to create any additional victims.

The very first thing to do when asked to fly search and rescue, aer-ial photography, or game spotting is therefore to very honestly eval-uate your personal capabilities, the aircraft's capabilities, and the weather. The most important skill to employ is not the canyon turn; it is the ability to say no.

- No, we can't go right now. We have to wait for better weather or more daylight.
- No, this aircraft can't handle that mission with an acceptable safety margin.
- No, as a pilot, I don't feel comfortable doing what is required.
- Or even yes, I can help, but I cannot go any lower than this in this much wind, or I cannot go any deeper into this particular canyon.

Don't worry about letting someone down. If the dangers are not acceptable, everyone involved would be ill served by an aborted mission, much less a crash. You will find that most people are quite willing to respect your judgment.

Rarely have I ever simply had to just say no to our local search and rescue coordinator Eric Berg, but I have often wished I were able to do just a little more and he has never complained that there are very real limits to what can be accomplished.

Game spotting.   If you feel a desire to take up this kind of flying, get into it *very* gradually. An excellent place to start is *game or livestock spotting*. The reason is that the animals you are looking for are active very early in the morning right after sunrise, when the air is normally very still. As the sun gets higher, the animals begin to lie up under cover and chew their cuds. About the time the weather typically becomes more turbulent, you can't see the animals anymore anyway. Fly these flights very conservatively. You will be amazed at how easy it is to spot game from perfectly safe altitudes, and how impossible it can be to spot anything when you are too close to the ground.

Air-to-ground photographs.   *Air photographs,* at least professional-quality air photographs, require you to be in the air at the moment the sun angle is right for your subject. Professional photographers refer to the hour around sunrise and the hour right around sunset as the magic hours. Fortunately for you, these are the less turbulent times of day. Unfortunately, some photographs require overhead light and must be taken at midday. Others require the view from the lee side of a ridge, etc.

In taking air photographs, while you are still governed by regulations that require you to be 1000 ft above a town, 500 ft above a built-up rural area, or 500 ft from any structure or person on the ground, it is perfectly legal to fly over and take photographs of things on the ground that the owners do not want to have photographed. Working for Lighthawk, we photograph polluters at will. Although we have agreed to respect his privacy, others locally will pick up an easy couple of hundred dollars from the tabloids with a photographic pass of Tom Cruise's house whenever he is in town with his children.

To get the wing, tail, landing gear, or propeller out of the shot, it is often necessary to perform a full-rudder deflection sideslip. Remember that in a slip your airspeed indicator is not reading correctly, and that stalling in a slip is the easiest way to induce a spin. You concentrate on flying the plane, not on the photograph. Keep the nose down when slipping.

A professional photographer will not take a picture through the Plexiglas window of the plane. She or he must be seated next to an openable window and, if right-handed, may need to be sitting in the left seat. For photographs using a wide-angle lens, I frequently remove one door of the aircraft and one of the front seats. The photographer then sits with his or her feet out the door and resting on the landing gear and is belted in with the existing seat belts. *It is not legal to do this unless your aircraft has an STC approving it!* You should also be aware that with a door removed, the performance of the aircraft changes significantly if you are in a slip to either direction. Drag goes way up when you sideslip toward the missing door.

Air-to-air photographs. Taking air-to-air photographs requires formation flying, something the average pilot never does at all, much less practices. There are schools that specialize in this kind of instruction. As with anything else, while you can teach yourself how to do it, you are better off seeking instruction. The fundamentals are that one plane is the lead, whose job it is to fly as smoothly as possible. The second plane flies formation on the first. Any move by the lead plane is called out ahead of time on the radio by the lead pilot so that the second pilot can anticipate the change. If the second aircraft is too far back, the pilot adds power until she or he is in position, backs off power to slow the extra speed, then adds back in power to hold position—three throttle moves.

Search and rescue. People have the unfortunate practice of getting in trouble in very inconvenient places at very inconvenient times. Note that this pilot is flying down the canyon toward lower terrain.
John kounis photo.

The lead plane never tries to do the actual formation flying. On the contrary, the lead concentrates strictly on flying very smoothly and predictably and lets the second plane handle the formation work entirely. Both of you doing this at the same time can be a nightmare of uncoordination.

By FAA regulation, you cannot carry passengers when you formation-fly. You must agree ahead of time on the ground what you intend to do and how you will do it. You agree ahead of time on the ground how close you have to be, how you will form up, and so on. One thing you must agree on is how to separate if the second aircraft loses sight of the lead. Typically, one aircraft will go up and the other down, and you turn away from each other. It is quite likely you will lose sight of each other if a lot of steep banked turns are required, the formation is tight, and the sun gets in your eyes.

The photographer directs the action. The photographer tells the lead pilot where to fly, then directs the second plane into position for the shot. While there are hand signals for this kind of thing, the use of the radio is much less ambiguous. Normally the photographer is in the lead plane taking photographs of the second plane, but you don't need to be restricted to that. By prior arrangement, the lead can be changed and the photographer's plane can follow the hero plane for a greater variety of shots.

There are two flight patterns used in over 90 percent of all air-to-air shots. Either the lead plane flies straight and level, putting the second plane in front of the appropriate scenery; or the lead plane flies in a circle with the second plane to the outside matching the bank angle. The photographer then shoots during the portion of the circle where the angle of the light is best. This latter method often produces a more dramatic shot because of the bank. It also allows the photographer more chances at the right background. Unfortunately, the best part of the circle for the picture is usually when the second plane is lit by direct sun, which means the second pilot is blinded while looking directly into the sun and is in the worst part of the turn to keep the distance and spacing correct.

As with air photographs of the ground, a professional photographer will not shoot through the plastic window of the plane. You must make provisions to open a window. When flying with Ron Kantor, we typically remove the photographer's seat and make provision to belt him in facing backward to shoot a plane in formation slightly behind us, to get the pilot's face in the shot.

It is easiest to fly formation with two identical aircraft. Nevertheless, we have had excellent results shooting twin turboprops in close formation with a Cessna 210. The difficulty lies in the fact that the 210 flying at near its maximum 170 knots is near the slow end of the control range for the turboprops. The professional pilots of the twins made it look much easier than it actually was. We had a very difficult time shooting a Beech Bonanza for a magazine cover from a Cessna 150 because the 150's maximum speed was simply too close to the Bonanza's stall speed. Although our Cessna 172 trainer is not exactly identical to the Cessna 185 we used for the camera plane, the overlap of speed ranges was great enough to make formation very easy, and the result was the cover of *Pilot Destinations Magazine*. The lesson here is that if the two planes you have available are too much of a mismatch, it will save you time and money to rent something more appropriate.

Search and rescue

The trouble with *search and rescue* is that the victims tend not to take your flying needs into account when they decide it is time to get lost

*Air-to-air photography. The camera plane flies the lead, everyone else
flies formation on him. All the pilots have considerable formation
experience, and they went over all the procedures before they took off.*
John kounis photo.

or injured. Time can be very critical, and you can easily find yourself
going out with night coming on or bad weather all too present.
Because the need to act quickly is critical, rushing into something
you shouldn't be doing can be the result.

Search and rescue flying falls into three broad and overlapping cat-
egories of search. Probably the first thing that happens is that the
local law enforcement search and rescue coordinator gets one plane
up as quickly as possible to get right to the search area as fast as possible. This kind of unplanned quick search can definitely save someone's life. Often, though, it won't. The odds are very much against a
quick search finding someone because time is short and so much
terrain has to be covered. Gratifyingly, the outcome of many of these
searches is that the victims find themselves and walk in, and the
search is canceled.

The next phase may or may not be a massive effort, often run by the
Civil Air Patrol. There may be multiple aircraft, and ground radio
coordination stations will be set up. Different people will be
assigned different areas to cover. Some aircraft will have special
equipment to direction-find emergency locator transmissions (ELTs).

The Air Force or National Guard may participate. This is a big, slow-moving, very cumbersome effort, but a much more thorough one.

Independent of either of these two phases is something like a road search for a missing person. I have been called upon to fly hundreds of miles of highway looking for missing motorists. These searches can be days after the motorist failed to turn up, and it is by no means clear that the motorist is actually overdue on the road, or even which road. A river trip a day overdue may or may not precipitate an air search of the canyon they were running.

In both these cases, you are flying somewhat low into an area where electric power lines can be very high. Power lines tend to cross rim to rim over canyons and valleys where your search area is well below the rim. A qualified observer who knows the location of some of these lines is highly desirable!

Plan.   Discuss your objectives before the flight. Agree on what you are going to do and where you are going to go. Check the map. Identify and remember hazards. Assess and discuss the weather, and agree beforehand that you will turn back if it goes bad. Then stick with that decision.

Coordinate.   If you will be working in conjunction with someone on the ground, or in conjunction with another aircraft, discuss plans together. Agree on radio frequencies and write them down. If there will be another aircraft working with you, discuss tasks, procedures, and flight plans. Be honest in admitting your limitations.

An observer.   You *cannot* effectively or safely fly the plane and look for things on the ground at the same time! You need a qualified spotter (or photographer) with you. It is the pilot's job to fly the plane and look out for terrain and other hazards. The spotter does the actual searching and directs the overall search pattern. If you need to coordinate with people on the ground, the spotter handles that radio traffic.

Every year game spotting and search and rescue produce accidents. The reasons seem to be twofold. First, the pilot is intent on looking for things on the ground, rather than flying. Second, this is a very much more dangerous kind of flying than it seems to be. Yet very low-time pilots who haven't flown for months suddenly get the idea that their hunting buddies are a great source of funds to share some of the costs of flying.

I find that I frequently have the luxury of flying with photographer Ron Kanter or search and rescue coordinators Eric Berg and Dean Gianpietro. They know how I fly, and I know how they like to be able to go about their business. They know the local area very well, and that makes it easier for them to direct me where they need me to go. Because we are well tuned in one another's needs, the flights are far less stressful and probably safer. Because they are very experienced at spotting things on the ground, they see things I would miss if I were doing the spotting, and because they are used to flying around in small planes, they are comfortable doing their jobs while letting me do mine. Once after about 45 min of extremely steep bank turns and chandelles under an overhanging cliff very, very deep in a canyon just wide enough to turn around after we had definitively located (and therefore saved the life of) a downed paraglider pilot, Eric turned to me and asked in a very relaxed voice if this was what I normally did all day long. No, usually I pretty much try to stay far enough away from the rocks not to scratch the paint. But Eric was not distracted in the least and was able to figure the whole problem out and set up the rescue. Inexperienced spotters can't find things on the ground they are looking for, and frequently can never really orient themselves and find where they are looking for.

Taking up amateur photographers to photograph their own ranches, for example, can mean nice pictures, but takes a lot more time and seldom produces the specific result they had hoped for. Going after the same photograph with Ron Kanter first involves many hours of research on his part, including using a relief model of the local terrain to predict the right sun angles. He then goes through a review of dozens of photographs similar to the one he hoped to bring back this time. Finally, once in the air, he draws on not just thousands of hours of pilot experience, but also thousands of hours of experience taking photographs from planes. The flight itself does not take that long, and the result is well worth the high prices he commands.

Fly just a little higher. Normally 200 or 300 ft in the air is low enough. Hotshot pilot that you are, you can undoubtedly fly right down in between the treetops; but when you do, the ground goes by too fast for the spotter to see anything, or the photographer to get anything but a blur.

In fact, many times flying photographic missions, we find ourselves well above 16,000 ft msl looking at subjects 7000 or 8000 ft below us. Many subjects look better shot from farther away with a long lens rather than close up with a wide lens. We don't have to mess with the dangers of low altitude at all.

It may be best to first scout the general area from, say, 1000 ft agl, then make passes at 300 ft, and finally make a check of some specific things you have already spotted with a very low pass.

Sideslips

Sideslips allow you to hold a desired heading while either picking up the wing tip or lowering a wheel to provide a better view. You can also move the wing forward or backward out of the field of view without turning.

However, these sorts of flights also have you low down, flying very slowly. A stall in a sideslip is the easiest way there is to enter a spin. Close to the ground, recovery is doubtful. Be cautious. In particular, avoid slipping in a nose-high attitude, and at least in single-engine Cessnas, avoid slipping with full flaps. An open window or removed door changes the behavior of a small plane in a slip. You can, and may have to, slip in this configuration. Just enter the slip very gradually and stay well above the stall speed.

A useful exercise for this kind of flying is to look at a landmark straight ahead of you on your next flight. First, lift the nose of the plane just above the landmark. Then, using the rudder, move the nose to the right, while correcting with the ailerons to keep the wings level. Then lower the nose, use the rudder to move it left of the landmark, and use the elevator to bring it back up. You can draw a perfect rectangle with the nose while holding the wings in zero bank.

Next look at the left wing tip. Move it forward with the rudder, drop it with the aileron, move it back with the rudder, and raise it again with the ailerons. You can draw a rectangle with the wing tip while holding constant-pitch attitude.

Descend canyons

If what you are looking for is in a canyon, it may well be all the way down in the very bottom of the canyon. Ideally, you will always fly *down* canyons. By entering the canyon from the top, you are already headed toward lower terrain. You are already executing your escape. If you are *headed toward lower terrain,* it is quite possible to get way down into a canyon if the situation requires it. If you are headed downhill, it is quite possible to fly under things such as bridges or power lines if the situation requires you to, in order to be low enough farther down the canyon. This can be extremely exciting and very heroic, but how often do you actually need to be that low? Flying up the canyon, you can get nowhere near that low because you would not have room to turn around.

If circumstances dictate that you must fly up a canyon, review the chapter on canyon turns. No matter how crucial the mission, you have to preserve the ability to turn back, just as in any other mountain flying. I pass along a bit of reasonable-sounding advice on this subject without comment: In a really narrow or really steep canyon, it might be a good idea to fly up the "bad" side of the canyon, that is, to fly up the sink or downdraft side. That way you will be aware sooner that you have reached the limit and must turn back. When you do turn back, your turn will take you out of sink into lift. Sometimes this is the best way, at other times not. Specific circumstances will dictate.

Bridges, wires, and arches

Remember one more time that you are getting down to the elevations of power lines and even other artificial things such as bridges. Colliding with any of these is probably going to be very serious. Power lines tend to be visible from some angles, but completely invisible from others. It is prudent to go over the map before the flight and see if there are utility lines across your route. Bridges have things hanging from them more often than you might think (particularly after the invention of bungee jumping).

It is not as difficult to fly under a bridge as most people think. Utah pilot Tim Martin said of flying through natural arches that the arch he intended to fly through was the same width as his runway, and he hit that every time. I guess he's right. No one has flown through more arches than Tim. But he would be the first to remind you that the consequences of missing could be almost certainly fatal. I am very firm in my belief that is it quite safe to fly through an arch with Tim, far less safe to try it on my own.

While it might be necessary to get down very deep in a canyon on a search, flying under the power lines is almost always an act of sheer bravado and unnecessarily dangerous at a time when all available rescuers are already committed to saving the person you are looking for.

Trim for climb

You can get very close to terrain beside you with reasonable safety. You cannot get close to terrain directly below you with much safety at all. If you *must* make a very low pass over flatter terrain, take a tip from crop dusters and military pilots. Trim the aircraft for a climb and hold it down with pressure on the yoke. That way if you are distracted, the aircraft will automatically climb away from the ground rather than hit it.

Look ahead

When you are flying close to the ground, as in landing, it is extremely difficult to accurately judge your height when looking down at the ground near you. Instead, do what the crop dusters do: Focus your vision way ahead toward the horizon.

Avoid heroics

Even if no one will admit it, the motive for a lot of this kind of flying is to be a hero. If that is your motivation, then you are in luck. Absolutely anything you do of this kind automatically makes you a hero! Heroism being a given, it can be achieved without resorting to heroics.

Fly in as conservative and professional a manner as the situation allows. You are providing an aerial platform, not a thrill ride. In fact, a plausible veneer of professionalism is sometimes even more impressive than pure heroics—people want very much to believe their safety is in professional hands!

The degree of cold you experience flying with the door off at high altitude in January can actually rise to the level of a safety consideration as well.

These types of flying might not sound all that hard to do. Doing them really well is very demanding, and the dangers involved are underappreciated. If you are inclined to try these things, work into

The acknowledged master pilot of the Utah backcountry, Tim Martin of La Salle Junction. Tim observes that this arch is as wide as his runway, and he hits that every time. Before you try the same thing, Tim's runway is narrower than his wing span. He is very careful with weather, and this was taken very early in the morning. Notice that Tim has put in about 10° of flap to slow down the plane a little and help tighten up his turn radius. It still requires an 80° bank turn to miss the canyon wall. Several years ago Tim was transporting a prisoner back for trial and flew through this arch. The prisoner asked the judge for, and received, time off from his sentence. Photo courtesy Tim Martin.

them very gradually under the guidance of someone experienced in this kind of flight.

Soaring flight

Pilots of small aircraft may benefit from some of the techniques that pilots of unpowered gliders employ. While a few sources of lift are at best esoteric and at worst useless to a small plane pilot, a couple can be routinely taken advantage of on any mountain flight.

Ridge lift

Ridge lift is the easiest kind of lift to envision and probably easiest to use in a powered aircraft. Wind blowing into the face of a ridge or any other obstruction is compressed together and deflected upward. A band of lift is formed along the upwind side of the ridge. The pilot surfs back and forth in this band of incoming rising air. The Wright brothers employed this form of lift in 1909 to keep one of their gliders airborne for 45 min at Kitty Hawk. At the time, their powered aircraft couldn't do much better.

The pilot of any small, single-engine plane can greatly improve the efficiency of the flight by planning a route that flies very close to and parallel to the upwind side of ridges. In fact, on a hot summer day with a load, it might be necessary to surf up a ridge to cross a high pass.

Thermal lift

Plumes or bubbles of rising warm air are called *thermals,* and circling up in these thermals is called *thermaling.* Although there are as many techniques for thermaling as there are pilots, the underlying principle is that you stay in lift by circling in the thermal. As soon as you start climbing, you immediately begin a sharp turn. Most sailplanes thermal at speeds between 45 and 60 knots, which is to say as slowly as they can possibly go without stalling. At those speeds, a turn with 45° to even 60° bank angle is still needed to stay in the thermal. Gliders flown aggressively in thermals are therefore very close to stalling in a steep turn, and a spin is the likely result. Stall/spin is the leading cause of sailplane accidents after the dangers of the tow itself.

If you want to try to thermal in a small plane, you slow down to your minimum sink rate, which is the same speed as V_{x}, your best climb angle. Unfortunately, this speed for even a small powered plane is still too high—your minimum turn radius will be wider than most thermals, so you will keep flying in and out of the thermal itself. Despite that, this form of lift can still be exploited advantageously. If you find you cannot climb high enough to safely cross a pass, fly back away from the pass and search until you find a location where your climb rate is better. Circle in that area until you are higher, and then make another try at the pass. Rather than try to absolutely core a thermal, you will probably achieve just as beneficial results from flying at best rate of climb V_y, and generally circling about where the thermals seem to be.

Once the thermals have climbed high enough to no longer be directly connected with their point of origin, they can acquire an internal energy of their own. When they climb to a certain altitude, the moisture in them begins to condense, and a small cumulus cloud is formed. Soaring aircraft will circle up directly under these clouds and then head out to circle up under the next cloud. Long lines of cumulus clouds are called *cloud streets* and are one of the principal mechanisms of soaring cross-country flight. In a powered aircraft, circling under each cloud is a waste of time; but you can certainly pitch the nose up and climb while you are in lift under the cloud, then pitch the nose down slightly and speed up to get through the sinking air in between.

Very strong midday convective lift produces turbulence, most of it right at the edges of strong thermals. Because they fly as slow as one-half of the speed of powered aircraft or less, and because of their long, thin, flexible wings, sailplanes do not feel this turbulence with anything like the same degree of severity. Also, they can avoid exposure to the turbulence by staying in the core of the thermal until they have climbed to an altitude where both lift and turbulence are weaker. As a powered aircraft pilot, you will have to decide if using the lift is worth the discomfort of the turbulence, or if picking a route in smoother but not lifting air is preferable.

Any lift. Whatever the exact source of lift, flying directly over the spine of a ridge is a likely place to encounter it. Flying over the center of a deep V-shaped valley or canyon is a likely place to encounter sink and turbulence.

While most of this text concerns itself with teaching you to avoid areas of strong sink, lift and sink go hand in hand. If you encounter strong sink, you can be sure that air is rising somewhere else. The puzzle for you is to find where the area of lift is and fly there instead.

8

More advanced mountain flying

At this point you know something more about mountain flying, but how do you go about getting better at it?

Practice

The obvious way to become more proficient, more comfortable, more capable, or just safer is to gain more experience. The military and NASA back in the early days of the space program in the 1960s tried a variety of things before choosing *gradated exposure* as a principal means of training. Gradated exposure simply means exposing someone to new or stressful things in small, gradated doses.

The theory behind this has both a practical and a psychological aspect.

From a psychological sense (oh great, the flight instructor is an amateur psychologist!), it is apparent that most people can be trained to operate efficiently under very high levels of stress. Consider just in the world of aviation such examples as Gordo Cooper quietly falling asleep while waiting for the ignition of the thousands of tons of volatile explosive rocket fuel behind his seat that would soon blast him into outer space. Consider that all branches of the military train their pilots to fly very professionally while people are shooting at them and trying to kill them. Consider the pilots of all the mountain commuter airlines routinely day in, day out flying difficult instrument approaches into mountain airports in icing conditions, or the freight haulers who do the same thing with a single pilot at night. Author Tom Wolfe invented the terminology. Pilots who can do this sort of thing have a certain indefinable something that sets them apart from the rest of humankind. Wolfe called it simply *The Right Stuff*.

I quote the relevant paragraph from Wolfe's book, *The Right Stuff* (Farrar, Straus, and Giroux, 1979):

As to what this ineffable quality was...well, it obviously involved bravery. But it was not bravery in the simple sense of being able to risk your life. The idea seemed to be that any fool could do that, if that was all that was required, just any fool could throw his life away in the process. No, the idea here (in the all enclosing fraternity) seemed to be that a man should have the ability to go up in a hurtling piece of machinery and put his hide on the line and then have the moxie, the reflexes, the experience, the coolness, to pull it back in the last yawning moment—and then go back up again the next day, and the next day, and every next day, even if the series should prove infinite—and ultimately, in its best expression, do so in a cause that means something to thousands, to a people, to humanity, to God.

Or consider you. Soon, if not already, you may fly a very small aircraft over very rugged terrain through unpredictable weather at the very limit of the aircraft's capabilities. Yes, you, in your awesome, roaring, fire-breathing hurtling piece of Cessna 172, Piper Cherokee, Husky, or whatever, are out there on the verge, rolling the dice with death! And you're bringing it right back again, time after time! You are somehow going to psychologically move yourself toward the true brotherhood of the right stuff! Presumably you are an average pilot. Now you want to become a mountain pilot, which requires more than average abilities. Can you do it?

Jumping in at the deep end probably isn't going to work for a number of reasons. You need a better method. Ideally, under stress the average person responds rapidly, exactly, and often automatically within the limits of his or her training, capabilities, and experience. Contrary to popular belief, most of us are hardwired to behave in exactly this way. We *all* have the right stuff. In popular jargon, this is called being focused. We *all* have this ability, but we all also realize that we don't always perform under stress just as we wish we would. Why not?

Some of the key words in the preceding paragraph are *training, capability,* and above all *experience.* You can't almost automatically, instinctively, do something you haven't been trained to do. You can't suddenly do something under stress that you aren't normally capable of doing in a low-stress situation. You can hardly be expected to do something automatically that you have never practiced. If you find yourself in a situation that demands you do all these things which you have never practiced or never been taught, you will certainly feel considerable anxiety, if not abject fear. As defined for this purpose, anxiety is the extreme worry brought upon you by extremely stressful situations. Anxiety is a considerable barrier to learning. *So don't try to learn that way!*

Here we are on very familiar ground. We are all aware of some of the more extreme behaviors caused by anxiety in extreme situations: screaming, running frantically in all directions, freezing like a deer caught in the headlights. Doing any of those things while flying an aircraft would not be terribly useful, nor would it impress your passengers very positively. But your internal makeup isn't oriented toward those responses. You are one of the true brothers. The way to ensure the true brother of the right stuff response every time is no more complicated than gaining more experience. The most effective method of using that experience to produce the desired response is *gradated exposure.*

Gradated exposure

The concept of applying gradated exposure is a simple one: Don't go right up on the worst possible day in barely adequate conditions. Choose to fly on absolutely optimal days, and then practice the things you may need to do later on another day under more demanding circumstances. If you practice things under very low stress, in a low-anxiety situation, then when a more stressful situation occurs, you have already worked out the flying techniques you will be using. Having these techniques somewhat ingrained means you will automatically find yourself focusing on them. You will also find that now doing these things requires less concentration, so you can direct more of your attention to other aspects of the situation.

By increasing your exposure to higher levels of stress in steps, very gradually, you automatically increase your stress tolerance.

Learning the technical aspects of flight works the same way: First you learn the simple things and concentrate on becoming very good at them. Next you add more elements, each building upon the things you already know how to do. Think back to the first time you drove an automobile at the age of 16. You could drive the car, but changing stations on the radio was so overwhelming that you almost had to pull over on the shoulder of the road to do it. If the wrong song was being played, you probably had to turn it off in order to remain focused on driving. But now you can simultaneously flip a new CD into the player and hit the particular track you are searching for, conduct important business deals on your cell phone, balance a lap full of fast food, drink a 64-oz Coke with the other hand, and turn around and yell at the kids in the back seat, all while merging with 80 mi/h traffic on a freeway on-ramp. Okay, maybe this is not exactly a good idea, but you see the principle. As a pilot, you can simultaneously fly the airplane, talk on the radio, lower the flaps, slip for a crosswind, and reduce the power. In unexpectedly bad weather, since you have practiced doing all those things, you can do them reasonably automatically while you adjust your flight path and speed as well as deal with the weather.

But presumably you are interested in adding some new skills as well as improving old ones.

Advanced mountain flight instruction

Naturally I recommend you fly with us at

Mountain Aviation Services
Telluride Regional Airport
37 Hillside Lane
Telluride, CO 81435
www.discovercolorado.com/mtnaviation

But we hardly have a monopoly on mountain flight instruction.

If you are based in a mountainous area, chances are that any flight instructor you work with has had some mountain experience. In Colorado you are really in luck because the Colorado Pilots Association (CPA) regularly conducts mountain flying courses. In fact, you are doubly in luck because many of the people taking these courses or attending CPA-sponsored seminars are in fact mountain flight instructors themselves. The Colorado Pilots Association has a mountain flying committee which regularly updates its programs.

Colorado Pilots Association
Box 200911
Denver, CO 80220
www.coloradopilotsassociation.com

If you plan to fly in some of the other mountain states, there are organizations whose primary focus these days is the fight to keep their runways open. (Let's face it, if you fly, you will eventually need somewhere to land.) As of this writing, they do not conduct a specific program like the CPA mountain flying program, but they can refer you to the most appropriately qualified instructors in their areas, and they certainly could use your support.

Utah Backcountry Pilots Association
C/o Steve Durtschi
792 South 150 West
Centerville, UT 84014

Montana Pilots Association
Back Country Airstrip Foundation
www.montanapilots.org

Idaho Aviation Association
Box 2010
McCall, ID 83638
www.flyidaho.org

Oregon Pilots Association
www.oregonpilots.org

Alaska is its own world. Remember back in Chapter 1 that Alaska always leads the rest of the nation in aviation accidents? But if you are going to Alaska, you also realize that a greater percentage of the population of Alaska relies on small aircraft for their basic daily needs than in any other region on earth. If you learned to fly in Alaska, you are already in touch with experts you need to consult about the specific needs of that kind of flying. If you are moving to Alaska, you will want considerable local guidance. For starters, contact:

Alaska Airman's Association
Box 241185
Anchorage, AK 99524
www.alaskaairmen.com

Southeast Alaska Pilots Association
1621 Tongass Ave., Suite 300
Ketchikan, AK 99901
Seapa@kpunet.net

Further reading

For small-scale meteorology:

Flight Conditions (Micrometeorology for Pilots)
Dennis Pagen, self-published
1184 Oneida St.
State College, PA 16801

Dennis has a number of outstanding books of considerable interest to all pilots of light aircraft.

For more on mountain flight:

Mountain Flying Bible
Sparky Imeson, self-published
Aurora Publications
Box 537
Jackson Hole, WY 83001-0573

Sparky Imeson is something of a guru on mountain flying and covers a lot of ground in a highly technical way.

The title tells it all:

How to Crash an Airplane (and Survive!)
Mick Wilson, self-published
Aviation Forum Company
Box 2885
Loveland, CO 80539-2885

For advanced maneuvers:

Basic Aerobatics
Szurovy and Goulian
McGraw-Hill
Two Penn Plaza
New York, NY 10121-2298

This book is a little technical for a pure layperson, including me, but worth the effort:

Mountain Meteorology—Fundamentals and Applications
David Whiteman
Oxford University Press
2001 Evans Road
Cary, NC 27513

This is much easier to grasp and contains very usable information:

Clouds of the World
Richard Scoria
Stackpole Books
Harrisburg, PA

This is out of print, but browse the used bookstore.

The title describes the subject. This is a more in-depth examination than can be found elsewhere.

Exploring the Monster—Mountain Lee Waves: The Aerial Elevator
Robert Whelan, self-published
Wind Canyon Books
Box 1445
Niceville, FL 32588

St. Exupery wrote several books of considerable literary merit. This one in particular has some excellent descriptions of flight in weather:

Wind, Sand, and Stars
Antoine de St. Exupery
Originally published in 1939, still in bookstores.

This is the definitive work on aviation psychology. You don't need to fly to read it. It later became an award-winning Hollywood film:

The Right Stuff
Tom Wolfe
Farrar, Straus, and Giroux Publishers
New York, NY

The following is quite likely the best book ever written about flying. People who never fly should read it to understand those who do. Mountain pilots should read in particular about icing over the Appalachians and scud-running approaches to upstate New York airports:

Fate Is the Hunter
Ernest K. Gann
Simon & Schuster, Touchstone Books
New York, NY

This is a personal account of learning to fly the mountains:

Ice Runway
Roy Mason
Beaufort Books
9 East 40th Street
New York, NY 10016

Written in 1924, this is an excellent account of early flight by a superb author:

Wind in the Wires
Duncan Grinnell-Milne
Doubleday & Company
Garden City, NY
Long out of print.

At least this book was modern in 1955! It contains excellent basic flying information. Try the used bookstore.

Modern Airmanship
Neil D. Van Sickle, USAF
D. Van Nostrand Company
Princeton, NJ

Jim Hurst is one of the pioneers of Utah backcountry flying; Pearl Baker is an historian with an intimate personal knowledge of the Colorado plateau. This is seldom seen in bookstores. Try phoning the Moab, Utah, airport for a copy.

Rim Flying Canyonlands with Jim Hurst
Pearl Baker
A to Z Printing
Riverside, CA

Three particularly relevant magazines for mountain pilots are as follows:

Southwest Aviator
3909 Central Avenue NE
Albuquerque, NM 87108
www.swaviator.com

Pilot Getaways
www.pilotgetaways.com

Northern Pilot Magazine
Box 220168
Anchorage, AK 99522-0168
www.northernpilot.com

These magazines can be found at the FBO of almost any western
U.S. airport.

Summary

Skip the whole book—just read this. Here are the basic dos and
don'ts for mountain flight:

Do

Do carry enough fuel to be able to change your plans en
route and go to alternate airports very far away. Despite the
added weight, the only time you have too much fuel is when
you are on fire. The most useless thing in an airplane is
fuel you didn't pump.

Do plan your route ahead of time, and mark it on the chart for
easy reference in flight.

Do calculate the expected time en route and arrival times.

Do file a flight plan, use flight following if your aircraft flies
high enough to use radar services and do remember to close
your flight plan.

Do take advantage of en route weather services.

Do check to see if your ELT batteries still hold a charge.

Do carry an emergency survival kit, including a first aid kit you
know how to use, a sleeping bag, warm clothes, food, and
matches.

Do fly early in the morning and get on the ground and stay
there by noon.

Do stay low in the valleys in winter. Winter winds aloft can be
extreme.

Do ask local pilots for advice!

Do obtain a weather briefing from Flight Service before departure,
and in particular, get winds aloft.

Do preflight thoroughly, because there are very few good emergency landing spots in the mountains.

Do calculate the takeoff distance from your operator's manual before takeoff, and do calculate climb rates for your load and altitude.

Do carefully set the engine's fuel mixture on the ground before takeoff, and do recheck it from time to time in flight. A too-rich engine will run poorly at altitude.

Do use all the available runway for takeoff.

Do climb at the best rate of climb speed, and get as high as possible as soon as possible.

Do allow at least 1000 ft of clearance over all terrain. Do allow more than that if it is practical to do so.

Do climb up the side of a valley, not the center. Do leave room to turn back.

Do cross passes and ridges at a 45° angle at least.

Do descend slowly in turbulence.

Do fly your normal indicated approach speeds on final approach.

Do seek additional mountain flight training.

Do respect the weather and your personal limitations.

Don't

Don't fly a small, low-powered aircraft IFR in the mountains.

Don't fly a small aircraft at night in the mountains.

Don't fly under ceilings of less than 2000 ft, even though clear of clouds is the legal minimum.

Don't fly with less than 10-mi visibility in the mountains, even though 1 mi is the legal minimum in G airspace, and 3 mi is the minimum in E airspace.

Don't fly a small aircraft in the mountains when ridge top winds exceed 20 knots or winds at 15,000 ft exceed 30 knots. These winds can produce downdrafts and turbulence far in excess of your aircraft's capabilities.

Don't—at least for most small, low-performance aircraft—fly with a load anywhere near the aircraft's certified maximum gross weight. In many low-performance aircraft at high altitude, 200 lb under gross weight might be a very reasonable maximum.

Don't run the engine at sea-level mixture settings.

Don't shock-cool the engine on a descent.

Don't ever feel that you have to get to your destination today. The weather will always be better tomorrow.

That's all there is to it. You don't have to go back and read the whole book unless you want to.

Just a very pleasant photograph of a beautiful evening. Aren't you glad you can fly? Gerrit Paulson photo courtesy SW Aviator Magazine.

Postscript

My first solo flight

Did you think that because I wrote this book, I must have done all sorts of things right that everyone else didn't? Do you think my judgment must be better than anyone else's? I didn't, and it isn't. I learned to be a reasonably competent pilot despite the way I went about it, not because of it. I learned to instruct in part because I was such a difficult student for my instructors that I really had to figure things out rather than just copy my instructors and do what they did. Take my first solo:

Most pilots will tell you that their first solo flight was fraught with anxiety, emotionally draining, but actually went quite well. They can thank their instructors for that. No one signed them off to solo until they spent an apprenticeship learning and demonstrating the skills to do it safely. The first solo flight was hardly the end of the instruction process, and getting a license was just the beginning of the learning process. In short, most people are guided through their first solo flight by someone who knows how to help them do it safely.

But you don't have to do it that way. The first aircraft I flew came in a cardboard box. No one signed me off to fly it.

Three of us were going to learn to fly together. The box contained a very used paraglider that one of my friends had traded from a visiting German kayak racer. Instructions were preprinted on the box, underlined several times in marking pen and followed by four exclamation marks. They read: **"Use no knives!!!!"** That plus "This side up," "Do not stack," and "Fragile" was the full extent of my written instructions. No doubt you have received similar boxes with similar instructions and similar little icons. But my friend had been given verbal instructions as well. He was told (with at least the same number of verbal exclamation marks and emphasis): **"Be very careful!!!!"**

Very carefully and without any knives we opened the box and pulled out the fabric within. We spread out the contents *this side up, without stacking,* and *carefully* amid the wildflowers on a steep, grassy hillside high in the mountains. Even to laypersons with no further instructions, it was obvious how the thing was intended to work. The wing was a big soft dacron sailcloth airfoil. Wind rushing in the openings in the leading edge would keep it inflated like a ram air parachute, only it was much bigger with a narrower chord and broader wing span. You would sit in a very rudimentary seat with a seat belt and crotch and shoulder straps. You would steer by pulling on toggles from which thin lines led up to the trailing edge of the wing. These could only be ailerons; or, more precisely, these could only be wing warping like the Wright brothers had used some 80 years earlier.

We knew you ran downhill until you just flew off into the sky. We even knew from a picture in a magazine that for some reason you ran while holding onto all the lines. (This is not exactly true, but it was a very small picture.) The big question was, How did you get the wing inflated and over your head? We hit upon a method. Two of us would hold all the fabric in our arms. The pilot would run down the steep hill as fast as possible. At just the instant the lines were about to go taut, we would throw the whole mess up in the air. My friend ran as fast as he could. We threw. He could no longer run down the precipitous slope fast enough and belly-flopped down the hill. We all tried. After a while there was no rush to be next. It was tiring and brutal. We were all bleeding. *Fragile!!!!* seemed pretty applicable.

But on my fourth or fifth turn, there was actually some resistance in the lines. The wing was way off to one side. It yanked me off my feet, I slid down the hill, and then suddenly I was flying through the air!

I couldn't possibly get into the seat. Instead, I dangled like a caught fish from the crotch straps of the harness. I had to start acting like a pilot quickly. I was in a gradual left turn toward the trees. Very, very gingerly pulling the left "aileron" control produced exactly the wrong result. I turned even more left and was definitely going to crash into the trees. For some reason, the "ailerons" were not working. I decided to turn even sharply left back into the hillside and land on the side hill instead of the tree. I pulled the other—the right—"aileron"…and miraculously turned right, not left, and flew out into blue space. (*This Side Up!!!!*)

So…they were brakes, not ailerons at all. The thing flew, but not at all in the way we expected it would.

But I was definitely flying! It was terrifying, but only mildly so. I think my biggest fear was that, for whatever reason, it would stop flying and plummet like a rock to the ground now hundreds of feet below. It was terrifying, but even more wonderful than terrifying. Despite the terror, I wanted to fly on forever.

Instead I came down. It was not too difficult, at least once I learned the ailerons were really brakes, to steer toward the center of a big, open field. Then I flexed my knees, kept my feet close together, and *slammed* into the ground with an impact that knocked all the wind out of me. (*Be Very Careful!!!!*) In seconds, my friends were all over me, ready to run back up the hill and try the same thing themselves. With practice we figured out a method of kiting the wing overhead without assistance. (*Do Not Stack!!!!*) After two or three flights each, we were far too bruised and bleeding to continue. The longest flights just barely exceeded a full minute.

Yes, not only had I taught myself to fly with no more instructions than *"Use No Knives!!!!"* and four exclamation marks, but I already knew all about it, so I no longer needed any instruction!

Was this a good idea? Was it in any way safe? Would I do it the same way again? Obviously not. Later I found some very competent instruction, and years after that I actually learned to pay attention to it.

But was it wonderful beyond all imagination? Absolutely! I would do it all over again in an instant, but I know enough now to do it under adult supervision from a qualified instructor.

Those first landings always hurt.

Better writers than I ever aspire to be have tried and failed to convey the wonders of flight. All my life I had lived in the mountains. I had climbed them, skied them, kayaked their rivers, run through their forests, and cross-country skied their valleys. I had camped in them year-round and chosen to live in a small log cabin in the woods to always be close. But there was a whole new world in the air around them—a world only just above the reach of my fingertips, but a world whose scope and depth I had never dreamed of.

And suddenly with a device resembling a big tent rain fly and some string, I was soaring high above the tallest mountains with nothing but thousands of feet of blue space between the soles of my sandals and the ground below.

I was a small child again. I could fly, and my life was never the same.

Index

About the Author

Fletcher Fairchild Anderson is the owner/operator of Mountain Aviation Services and a multi-rated flight instructor, charter pilot, and corporate pilot. He has given more than 2000 hours of mountain flying instruction; those sessions led to the development of a tutorial for students, upon which this book is based.